HUXTABLE

EAST BUCKLAND

THE STORY OF A

NORTH DEVON FARM

AND ITS NEIGHBOURS

Edited and Published by

Barbara Payne

Huxtable Farm, West Buckland, North Devon. EX32 0SR

www.huxtablefarm.co.uk

British Library Cataloguing-in-Publication Data.
A catalogue record for this book is available
from the British Library.

Cover Design by Jacqueline Payne

ISBN 978-095210550-3
Printed in Great Britain by
Arthur H. Stockwell Ltd
Torrs Park   Ilfracombe
Devon

CONTENTS

INTRODUCTION

"*Can you tell me more about........?*" This question, which is so often asked concerning people, places or properties, is particularly applicable to "Huxtable" and its neighbours. It is in response to such questions that this book has been written. "Huxtable", a Hall house, was built in the early 16th. century and the present farm is of similar size although no longer home to the descendants of the Huxtables who were here early in the 14th. century. These two features alone make "Huxtable" a unique and very interesting farm.

In the book every effort has been made, with the help of the Record Offices in North Devon and Exeter, to verify factual statements although the chapters are the personal contributions of the individual writers.

The book starts with a chronological account of the living history of the farm and is the result of a local study by Barbara Payne.

It is followed by a genealogy of the Huxtable family which is considered to have originated here in the early 14th century. This chapter has been written by John Huxtable, whose family is from this area, and who has carried out the family research over many years. An appendix gives details of this, which may be of interest to anyone investigating their own genealogy.

Next there is a report, with plans, of the archaeology and architecture of the buildings. This has been prepared by John Thorp of Keystone Historic Buildings Consultants.

The major influence on both the area and on Huxtable is the Fortescue family. In 1454 Martin Fortescue was the first Fortescue at Castle Hill, and from 1797 the family were landlords of "Huxtable" for nearly two hundred years. Lady Margaret Fortescue has kindly written of her family in this context.

Although Lady Margaret's family may be a dominating influence in this part of North Devon, West Buckland School, the immediate neighbour to Huxtable, now has equal, if not greater, prominence. There is a very close connection between the Fortescue family and the school for it was founded by them in 1858. Mary Cameron, who has written other local studies, and whose husband is a teacher at the school, has written Chapter 5, because many of the "*Tell me about........?*" questions refer to the school whose physical presence dominates the area.

Having established the setting to "Huxtable", Doreen Ridd next gives her personal account of the years that the Ridd family lived at the farm, years that included World War II. We are most grateful for these chapters written in memory of her husband Fred. It was from the Ridds that the Paynes bought the farm in 1980.

Finally there are the chapters concerning the farm in the years of its ownership by the Payne family. Many changes have taken place, in the main part due to the influence of Antony Payne and his wife Jackie who has written a large part of the concluding chapters. We hope you find this book interesting and that you enjoy reading it as much as we have all enjoyed writing it.

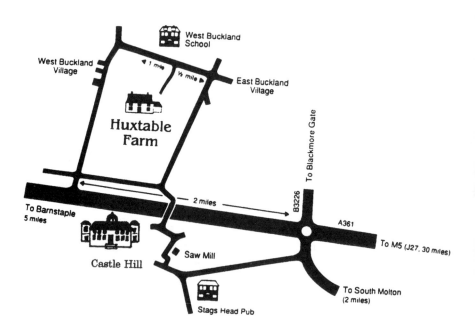

Huxtable Farm and its neighbours

Chapter 1. THE HISTORY OF HUXTABLE FARM

THE ORIGIN OF THE SURNAME "HUXTABLE" (1330-1711).

"Huxtable" is a small, 80-acre working farm some 650 ft. above sea level and situated on the edge of Exmoor in the parish of East Buckland, North Devon. The name Huxtable is first noted in the Lay Subsidy Rolls of 1330.[1] Here a *John de Hokestaple* is referred to, the name meaning "a spur of land with a post". This spur of land or hook can be seen on the present day ordnance map. From the early 14th century onwards the name "*Hokestaple*" only appears in the records of East Buckland until such a time as other "Huxtables" are mentioned in the adjoining parish of Charles and, later still, in the nearby parishes of Challacombe, Bratton Fleming and Berrynarbor. There are changes in the spelling of the name, by 1402 there is a reference to *John Hosestaple*.[2] Later again, in 1524, the name has changed to *Hucstapull* and in 1540 to *Hukestabull*.[3] From this it can be assumed that where a distinctive surname is found over centuries in one place in successive taxation returns and the like, family continuity can be presumed even if the descent cannot be proved step by step.

Much of this early history has been relayed to me by a John Huxtable. I am indebted to him for his help which generated in both of us the determination to trace the occupants of "Huxtable" for the period 1674-1711. We already know that the Devon Subsidy Rolls 1524-7 show only one *Thos. Hucstapull* in Est (sic) Buckland.[4] The Muster Roll of 1569 shows in East Bucklande Parrishe (sic), "Archer Richard Huxtable" and the Hearth Tax of 1674 again only one Huxtable in East Buckland. ("Thomas Huxtable" 5 Hearths). There is a discrepancy here as the present farmhouse has 3 hearths with traces of only one other earlier. It seems safe to assume that the occupants of Huxtable Farm were Huxtable at least until 1674, and that the will of one Thomas (11th July, 1622) that appears in Appendix 4 refers to one of them. In Chapter 2 John Huxtable gives more details of his work on the family genealogy, whereas this chapter continues with the history of Huxtable farmhouse.

THE EARLIEST LEGAL DOCUMENTS AND THE PURCHASE OF "HUXTABLE" BY THE EARL OF FORTESCUE (1711-1797)

An interesting house with an interesting family as the first occupants. But was there a house before the early 16th century? If not, where had the first Huxtable lived? Whether they were there or not, (prior to the early 14th century) may never be known although East Buckland (Bochelando) is referred to in the Domesday Book. From this we learn of three different holdings:- Wulfmer (24 acres), Ulf (6 acres) and Aldchuil (52 acres). This totals 82 acres which is interesting as the present parish is only 1,884 acres. This total includes much neighbouring land listed in separate farms in the Domesday Book. Neighbouring farms were also mentioned in the Domesday Book so we can conclude that it was a well populated and cultivated area in the 11th century.

The only deeds of the farm now held by the owner are of its sale by Lady Margaret Fortescue in 1975 and by the Ridds in 1981. Early Fortescue papers were, therefore, researched at the Devon Record Office in Exeter where they had been deposited by the family. The records traced through this collection commence in 1711 when a certain Richard Nott of Swimbridge, son of an Exeter merchant, sold "Huxtable" (which was now only 1/6 of his messuage) to James Buckingham as part of a marriage settlement to S. Burgess who was from the adjoining parish of West Buckland and whose marriage portion was £250.

At the Devon Record Office documents were seen concerning leases and re-leases, marriage articles and settlements, mortgages, lands, re-lease of legacies under will, fines and a covenant to levy fines and from these it is possible to trace the ownership by Buckingham up to 1767. By that time there were problems, the property was mortgaged and also rented out at a "peppercorn" rent (to a Joseph Simobar) until it was bequeathed in 1777 to Hill, son-in-law of Buckingham. Unfortunately this will has not been traced. (During War II Exeter was bombed and many records, including wills, were destroyed.)

These documents are not easy to read because of the older form of language, legal jargon and the actual state of the records. However, from these documents in Exeter it was possible to trace the ownership from 1777 to 1797 when "Huxtable" was sold to the Fortescues by the Reverend Hill and Metherall and Chapell, all descendants of Buckingham. Confirmation was sought in public records, and The Land Tax assessments of 1780 showed that the ownership of "Huxtable" was transferred from Chappell to the Reverend Hill and then later, 1782-1787, these Land Tax assessments show that the owner was the same Reverend Hill whilst the occupier was John Passmore. Early parish records (1684-1836) also held in the North Devon Record Office, confirm that many Buckinghams were baptised, married and buried in East Buckland Church or churchyard. Similarly the name Passmore was observed. (There are also many Huxtables although it must be noted that none say "of Huxtable") It is, therefore, not possible to say who lived at "Huxtable" between 1767 and 1780 although we know who owned it.

4

The next, and final, sale of that period was on the 9th of January in 1797 when Hill (on behalf of Chappell to whom he was married) sold "Huxtable" to Hugh, the Earl Fortescue, for £50 plus £1,600. This raises another question for it is difficult to believe that 80 acres was worth that amount in 1797 (we learn later from the Tithe apportionments that no tithes were levied on "Huxtable", assessed at £32, and this may be part of the explanation). This document has not been fully translated nor has more been learned about the first recorded owner, Richard Nott.

THE FORTESCUES AS LANDLORDS (to the mid-1830s)

When Hugh Fortescue purchased "Huxtable" in 1797 for the, then, large sum of £1,650, John Passmore was the "sitting" tenant (in 1794 there was a rental agreement - Passmore to Hill - for £57). With the Fortescue papers in Exeter Record Office was a notice to quit to Passmore in 1807 and there is a discrepancy here in that the land tax assessments of 1798 show Thomas Headen (sic) as the occupier. On 25th March 1815 there was the first indenture of £85 between Thomas Heddon and the Earl of Fortescue and seven years later (1822) a second indenture of only £65. This reduction in rent would coincide with the agricultural depression during the first part of the nineteenth century and the trade slumps, 1819-20 and 1826. After the French wars, prices dropped, and from 1815 onwards times were difficult for farmers. There was a high poor rate and general unrest amongst farm labourers, many of whom were revolting against the allowance systems of poor relief. This too was the time of the Corn Laws (1815). Following a bumper harvest in 1813 (although not applicable to this particular farm) resentment at the Enclosure Acts added to the dissatisfaction in agriculture felt throughout the country. There was a poor harvest in 1829 and in 1830 "the last revolt of rural poor, ricks burnt and threshing machines broken" followed in 1833 by the "Tolpuddle Martyrs". Despite this we know, from the Land Tax assessments, that Thomas Heddon was still at "Huxtable" in 1831. From the Parish Register we also know of many Heddons in East Buckland from as early as 1684 to as late as 1821, when a birth to Thomas and Mary was recorded. (This register is still kept in East Buckland Church).

At some time prior to 1841 Thòmas Heddon left "Huxtable". A search was made via the rent books and surveys of the Fortescue Estate. These are not held in the Exeter Record Office, but in a warehouse annexe, and, when checked, none of these pinpointed the exact year when the tenancy changed although the undated rent assessment gives a good idea of the poor state of the land at the period when the tenancy was valued at £52-2-10. (See Appendix). In 1807 it was £84-8-6.

THE FIRST SLADER AT HUXTABLE

With the agricultural economy in such a poor state, it is fairly easy to understand why Thomas Heddon, with his young son, should seek a new life in America where he would hope to be his "own man" rather than a tenant farmer. We know he was followed at Huxtable Farm by the Slader family, three generations, who possibly came from the nearby village of North Molton. We know, from the 1841 census that there was grandfather, "aged", then John and his wife Mary (who was born in North Molton). The tenancy was presumably in John's name and they had, in 1841, four sons and three daughters. Up to 1834 (Thomas was baptised on 10th August 1834) there was no mention of Slader in any of the parish registers. In addition to the family, there were two living-in servants, so "Huxtable" must have housed twelve. From this it is safe to assume that Thomas Heddon did leave in 1833 as his great-grandson stated.

What happened at "Huxtable" in the next twenty years is uncertain. In White's Directory of 1850, John is still the farmer, as he is in the census of 1861. But by this time there is only one son left at home - Thomas - now aged twenty-two. John and Mary are in their later sixties and are, in fact, to die in 1862 and 1865 respectively. From the Parish Register we learn that the eldest son, John, died in 1844 aged twenty-three (but only eighteen by the census). In the churchyard of St. Michael's, the parish church of East Buckland, no gravestone could be found for John whilst there were gravestones for the two youngest girls who died in 1852 aged twenty-four and fourteen (but no parish burial record of the youngest). There is no mention of the eldest daughter, Joan, marrying in the parish and, at this stage, we do not know what happened to the remaining three children. If, indeed, six out of the seven children died in this period it is hard to imagine how the family overcame these personal tragedies. It was the period of the serious cholera epidemic,(1848-49) and of the Crimean War (1854-56) but more research would be needed to ascertain the cause of these deaths. At the time of the 1861 census, the elderly Sladers, with their one son, were employing three servants and at that time farming, for the farmer rather than the labourer, was entering "the golden age" of high farming methods (1850-70's) and conditions were improving certainly in comparison to the thirty years following Waterloo. The Royal Agricultural Society had been founded in 1838 and although many of the changes were in the east of England, this period prior to 1873 saw many improvements in agriculture. There were government laws for farm improvements and the advent of the railway linking Barnstaple to London. The railway was between Barum (Barnstaple) and Taunton. This must have helped the farmers in this part of Devon. (Huxtable Farm was only two miles from the nearest station of Filleigh). By 1862 John Slader had died (confirmed both by the Parish Register and a gravestone) and the 1866 Post Office Directory of Devonshire gives Thomas Slader as the farmer. By the 1871 census we know Thomas had married and had already three children. The family were still employing two servants. Three of the children were baptised in East Buckland but John - who was take over the farm - was not. A year later a fourth child (a girl) was born who only lived to the age of five. Yet another girl was to be born who also died aged only five years. It seems reasonable to ask: did the Slader children inherit some congenital disease? The other families referred to in this study (Huxtable, Buckingham, Heddon) all appear to have been both prolific and long living.

LATE 19th CENTURY - EARLY 20th CENTURY

Between these dates we learn little more of the family from such records as are available. Thomas is quoted as the farmer at Huxtable in two directories of 1878. The first of Thomas's children died, Emily who was born in 1872 and died in 1877. It was during this period that there was the third depression of the nineteenth century which was so severe that there was a Royal Commission called to examine the causes (1886). Agriculture was in a particularly poor state. In 1877 farmers were paid compensation for slaughtering diseased livestock - both sheep and cattle - and, in 1879, the weather was so bad that the harvest was described as the worst of the century.

In addition imported food from America and Australia had adversely affected the home markets. In contrast to this depression there had been many improvements, new Public Health Acts, extending the power of local authorities, Education Acts, making schooling compulsory and from 1871 there were changes in the Poor Law and many reforms due to Lloyd George and the Liberal Government of 1906-11. At a local level the North Devon Journal of 1883 published an article supporting the view that a seventy-acre farm, such as "Huxtable", should be profitable. (See Appendix 6). However, even more relevant are comments made by Jon Edmunds in *A History of West Buckland School*. This school (founded in 1860) adjoins Huxtable Farm and Jon Edmunds writes vividly of the bad weather conditions of this area - up to twenty inches of rain per year, cold east winds in the winter and the whole area cut off for weeks if there was snow. In 1881 there was the heaviest snow since 1814 and again the great blizzard of 1891. This information was recorded in the school magazine. From the census of 1881 we know that Thomas Slader was at "Huxtable" with four children, the youngest - Alice - was described as "weak and feeble" (she died, as confirmed by both gravestone and Parish Register, in 1886) and he is employing only one servant. In the census of 1891 the family is the same, Thomas and his wife Mary, their unmarried daughter Mary and two sons, John and Fred. These three unmarried children were all in their early 20's. In 1893 his wife Mary died (aged 57), but it is not until after World War I that we learn from the directories (1919) that John has taken over the farm (when he would have been over fifty). From discussions with local residents, and notes in the Parish Register upon their deaths, we learn that the eldest (and only surviving) daughter and younger son went to the nearby town of South Molton where they ran The King's Arms public house. It would appear, therefore, that John - as yet unmarried - and his father lived alone at the farm until the marriage between Jack (John) Slader and Elizabeth Seldom which took place in 1911 when both were middle-aged. Thomas was still alive but then 75 years of age and his burial was not registered in the parishes. Supposing this to be correct, then Huxtable Farm was managed during World War I by one middle-aged man. The last date mentioned (in Kelly's Directory) where Thomas was the farmer, was 1906 when he would have been seventy. There is no gravestone for him nor entry in the Parish Register. Huxtable Farm is not easy land, the soil depth is poor, and approximately twenty acres is steep, rocky "cleave" land where broom, holly and gorse must have flourished. A depressing "inheritance" for a single tenant farmer especially at a time when England was becoming increasingly urbanised.

THE LAST OF THE SLADERS (-1941)

John Slader, called Jack, is still well remembered in the village although he died, aged seventy-four, in 1942 (from both the Parish Register and graveyard). For much of the following information I am indebted to Mr Ridler, who, as a boy worked for Jack's neighbour, Edward Cotsford, who wrote *A Local History of East Buckland*, and Fred Dallyn, a neighbour. Long interviews, which were microtaped, were held and from them a sad story emerges. This pre-World War II period also covers the depression of the 1930s when Jack was already in his fifties and trying to farm "Huxtable" on his own.

Everyone I spoke to, who either remembers him or has learnt of him through those who knew him well, tells the same story. A childless man who married late, had no contact with his only remaining sister and brother (also childless and unmarried) and who in later life was crippled with arthritis. His wife, Elizabeth, was universally disliked (she was Elizabeth Seldom of nearby Swimbridge, a Methodist, aged over thirty). She is described as "evil" and "hard 'earted" and "bossed" the farmhouse. In the main room, the original hall, she kept hens - the sawdust on the floor being changed infrequently. If there was a fire in the hearth then she would remove some of the faggots brought in by Jack if she thought the blaze too high. The floor in a second room was embedded with jagged glass to keep the rats out. Any cheques received she banked in her own name and was described as always filthy and wearing men's heavy boots. She was to outlive Jack by twelve years and died aged eighty-four. Jack too was described as filthy, never walking but going everywhere by horse and cart. Sad, too, was the state of the farm itself. The lane approaching it was impossible to drive down, the land covered with bramble, gorse, holly and over-run with rabbits. No produce grew and the few sheep ran wild over the entire farm, frequently breaking into neighbouring land as the dividing hedges were left untended. In 1939 Jack Slader was still at "Huxtable", already over seventy, ill and by now in bed, but still the tenant of the Earl of Fortescue. (There is much local talk of the inefficiency of his agent). In 1940 the Ministry of Agriculture compelled farmers to grow a certain acreage of corn and this was only achieved at "Huxtable" by the neighbouring farmers ploughing and seeding the land. However, the state of the land was such that no crop grew and on Lady Day in 1940 Jack Slader was evicted and was literally "carted off" the farm. A sad end to nearly one hundred years of Sladers as tenants of the Fortescue family. One wonders why the Fortescues had allowed the farm to deteriorate to such an extent.

THE WAR, AND THE RIDDS 1940-75

What now follows is based on information supplied by the Ridd family, their neighbours and by the present agent of Lady Margaret Fortescue, Mr. Hugh Thomas. As already recorded, the farm was in an appalling state when Fred Ridd (aged 59), his wife and son, Fred (aged 24) arrived from Dulverton. (They had previously lived in the village of West Buckland where, in Leary Bottom, Fred junior was born). It was so bad that the furniture lorry was unable to get to the farm. The rent was fixed at £75 per annum, the local feeling being that it should have been rent-free whilst the farm was in such a bad state. The Ridds worked hard, aided by the arrival of a landgirl, Doreen Cass, in May 1943. On Boxing Day 1944 Doreen married Fred and two daughters were born. These three generations lived at "Huxtable" until Fred senior died of a painful facial cancer in August 1948 aged 67. Conditions were hard. There was no electricity, the ram that pumped the water was always breaking down and the senior Ridds were a formidable couple. Living conditions with a primitive kitchen and no bathroom were not easy for the young Ridds who received no proper wages and it must have been difficult for them in the period prior to Fred senior's death of such a prurient cancer.

By the 1950s when they were the tenants, there was help from a grant - the livestock rearing scheme in 1955. The lane approaching the farm was made up, a Rayburn and the bathroom installed and mains water laid on. There was electricity in 1960 and in the same year the thatch was removed and the roof tiled. In 1970 the Fortescue family had to sell many of their estate farms and much land to pay death duties, but it was not until 1975 that Fred Ridd was given the opportunity to buy Huxtable Farm and the only conditions to the sale were that the Fortescues retained the hunting and shooting rights. Therefore July 1975 ended the period of ownership by the Fortescue family, nearly 200 years in the history of a farm bought as an investment and sold to help pay off death duties. Doreen Ridd has written Chapters 6 and 7 giving her personal account of this period.

Now, once again, a three generation family lives at "Huxtable". Despite having the largest amount of stock ever to be farmed here, it is impossible for such a small farm, on marginal land, to be economic. Farm diversification is the current trend and E.E.C. grants for farm tourism have meant that Huxtable Farm is once again able to support its owners, although many of the original farm buildings have been converted for the late twentieth century style of farming, and house greater numbers than ever before. Many of the visitors who come as a result of this farm tourism show great interest in the history of this lovely old farmhouse and it is for them and for my family that this book has been written.

Notes

1. 17 Lay subsidies were levied between 1290 and 1334 and in 1332 *Est Bokland* had 10 people listed. This roll is the most complete published record of the period for Devon - it concerns the payment on moveable goods to the Exchequer.

2. The card index system in the Devon Studies section at Exeter library states "May 10th 1402 (3 Henry IV-Pt.II) Westminster. *John Hokestaple*, for not appearing to answer Alan Heddon touching a trespass". N.B. the card index also states "pardoned of outlawry......".

3. 1540 (circa) *Hukestabull John*: - monastery prisimir (sic) employed at East Buckland by Elinora Peckham, widow. This record occurs in lists relating to persons ejected from religious houses.

4. Also four more Huxtables in the adjoining parishes of Charles and West Down with alternative spellings - *Hucstapyll* and *Huxstapull*.

Chapter 2. THE HISTORY OF THE HUXTABLE FAMILY

Reflections of John Huxtable, a Civil Engineer whose roots are in the North Devon soil.

I know, because of my name and the few records that remain after enemy action in Exeter in 1942, that my roots go back through many centuries of farmers in North Devon. Whilst our direct line is not yet traceable beyond the 18th century, the development of families of our name deviates little from North Devon in general terms.

In the *Devon and Cornwall Notes and Queries* of 1965-67 Mr C. Whybrow gave a comprehensive up-to-date history of the Huxtables of Swincombe. He was of the opinion that the origins of the family stemmed from East Buckland parish, more precisely from Huxtable Farm in that parish. He said that one only had to look at the telephone directory to see what a heavy concentration of that name was within about fifteen miles of the original family farm. The name is found in all parts of Great Britain today but not in concentrations such as in Devon. In the "new" countries such as Australia, Tasmania, Canada, South Africa and the U.S.A. one finds the name, sometimes in small groups but, where I have had contacts with Huxtables in such places, in no case has there been a suggestion that their roots were anywhere but in Devon and West Somerset. Of course the younger generation will have homelands in all sorts of places and unless they enquire deeply they may never know where the old name originated or where it was most prevalent.

It is very difficult, in the absence of facts, to make categorical statements but, by going back over old family conversations, by visiting family farms and by adding what has been researched, one gains a sixth sense of the way in which origin and drift could have taken place.

Before the advent of steam everything had to be done by hand and horse-power. There were only tracks affording very little mobility of movement out of the area. People were virtually captive. Even steam was for static use, mainly for threshing and sawmills until the railways were built. Farming, especially hill-farming, was really hard going and not until the coming of oil and the tractor was there any relief. It is still tough going by most accounts, including my own, when after a lifetime of civil engineering and a wartime of military engineering, my son and I had a small farm, both returning, in some way, to our roots.

The 19th century saw really great changes. Large families had a chance to move away "aided and abetted" by Brunel, the Stephensons, Telford and others, and then came the dispersal of large farming families. My family occupied Ruggaton Farm and Bowden Farm (where John Jewell, Bishop of Salisbury during the 16th century was born, and where there is an interesting 15th century screen) in the parish of Berrynarbour, North Devon. They also lived at Wallover Barton (also called Walliford Barton and later Wallover Farm), Challacombe, North Devon. This farm was occupied by one of the sons of Great-Great Grandfather Huxtable of Ruggaton Farm. Two other "products" of Ruggaton Farm moved to a farm near Dulverton and thence to two farms in Somerset, now reduced to one holding only, viz: Woodlands Farm, Holford and Pathe Farm, Othery thence to Clevedon.

My maternal grandfather, the son of a Cumberland farmer, John Morton (Gale Hall, Melmerby), had a long-established family history. My grandfather moved to Tavistock with three others between 1841 and 1851 to manage an estate and later he had his own dairy in Launceston, N. Cornwall, where he had a family of nine of whom two emigrated to the U.S.A. and one to Canada at the turn of the century. None of the family went on to the land. I believe that this brief saga may be typical of the large farming families of the past 150 years, thereby in great contrast to the apparently static state of affairs for several hundred years before.

Those of us who have researched the name of Huxtable believe that the first known mention was in the Subsidy Rolls of 1330 when there was one *John de Hokestaple* (vide: *Devon Transactions* - C.S. Spiegelhalter B.Sc.) Then there is a tremendous gap when the name changed to various spellings. The present-day spelling, of which I know no variant, may have been the same in some places as far back as the 16th century. From the attached list of wills there are several different spellings but these may have been due to inability to read or write and therefore based on pronunciation.

Obviously, I have covered only one modern branch holding the surname and I hope that these few reflections may inspire others to dig deeply and help in building a more accurate record.

Amongst physical changes which I have seen are: no rick-yards (the combine-harvester has displaced the threshing machine and harvester); there are few, if any, faggot ricks, few hedgerows which need laying, no regular rabbiting (especially Boxing Day), few hayricks (silage-making having taken over), few cider-presses (orchards grubbed out) and no village smithy now, with its constant din, no stables and the noble cart-horse. Missing too: milking by hand and in the dairy, the time and effort taken to wind the separator up to maximum speed with its never-to-be-forgotten high-pitched sound. Perhaps, where animals are concerned and in spite of intense farming methods, eternal vigilance day and night, seven days a week, is still the norm.

Finally, in Appendices, there are:

Appendix 1. The Devon Muster Roll 1569
Appendix 2. Devon Protestation Returns 1641
Appendix 3. Gravestone, West Buckland Church
Appendix 4. A will of Thomas Huxtable 1622
Appendix 5. Archival notes

The Barn after conversion

Chapter 3. THE ARCHAEOLOGY OF THE FARMHOUSE

THE BRIEF

Keystone Historic Buildings Consultants was contracted by the owner, Mrs B. Payne, to provide a report on the Listed Grade II farmhouse at East Buckland, Devon (O.S. Ref. SS 664 308) from an historical and archaeological point of view. An analysis of the building and its structural development form the content of this report.

SETTING

Huxtable Farm is situated in rolling hilly countryside less than 10km southwest of Exmoor. It lies at the western end of the parish, closer in fact to West Buckland rather than East Buckland. It is built on the springline in a dip in the lea (northeast side) of a ridge between two valleys which carry streams southwards to the River Bray.

The main drive approaches the farmstead from the northwest. The first impression is of a modest farm of late eighteenth century or nineteenth century buildings but some of the farm buildings are earlier and the house has late medieval origins. Some of the farmbuildings have been converted but the general impression is still one of a traditional North Devon farmstead. The farm is on a gentle north-facing slope and has a loose courtyard layout. The farmhouse faces south onto the farmyard with a former shippen built onto its east end. A short distance to west of the house is a small building, formerly the pigsties. There is a stable block down the west side of the yard, and a 2-storey building of uncertain function (maybe a granary over a cartshed) along the east side. On the northern side there is a good bank barn terraced into the hillslope. The barn itself had wide opposing central doorways to the threshing floor and a (possibly secondary) horse engine house projecting from the north wall to west of the doorway. The basement has three doorways onto the former yard and probably contained a root store and shippen.

The farm buildings are mostly built of local stone rubble but the stable and the possible granary/cartshed include some cob. Very little original carpentry remains in any of the buildings which makes dating them particularly difficult. They are most likely late eighteenth century early nineteenth century in date (using older cob walls in a couple of places) but the plain style of their surviving detail could date from any time from the late eighteenth century to the late nineteenth century. It is unfortunate that the farm was exempted from the 1840's tithe assessment, when it was part of the

The Tudor archway leading to the inner room

Fortescue Estate, which would show (or not) the buildings on a map.

THE FARMHOUSE

The farmhouse is built across the slope on a rough east-west axis and faces south. It appears to be built of local stone rubble but the older parts are cob (the plastered sections on the west end and rear (north) walls. It has a recent slate roof which was rebuilt higher than the original thatch roof. The front has a 3-window front of twentieth century timber casements arranged with a loose symmetry around the front doorway, a little to right of centre. There is no evidence of the great age of the house from the exterior. It is however, a house with a long and complex structural history.

The present house layout is essentially that of a comfortable Georgian farmhouse in vernacular style. It has a basic three-room plan with the front doorway to an entrance lobby containing the main stair between the two principal rooms. At the right (east) end is a spacious parlour (now the sitting room) with a gable-end stack serving fireplaces in the parlour and bedchamber above. A narrow room is partitioned off from the parlour along the back of the building and once had a connecting door to the parlour. It was probably a buttery for storing barrels of cider. The central room (now the dining room) was the kitchen and living room. It has a large fireplace in an axial stack backing onto the bakehouse/service room beyond. The fireplace is built of local stone ashlar with an attractive segmental arch and contains a typical North Devon cloam oven behind a red brick doorway in the left hand cheek. There are two doorways through the rear wall into a lean-to outshut (the present kitchen) which was probably used formerly for pantry, dairy and other storage uses. The small left (west) end room was the bakehouse; its front lateral stack is missing its chimneyshaft but still contains a brick fireplace with a plain timber lintel and a massive side-oven. This room (now a second dining area) also contains the service stair rising against the rear wall.

This then was the Georgian farmhouse but it was the result of a large-scale refurbishment of an earlier house. It was a major programme of works; all the chimneystacks date from this phase, the whole of the front wall was rebuilt between 3-600mm (one or two feet) forward from the line of the original front wall and, as far as can be seen, the parlour end section is a complete rebuild. The most likely time for such major alterations was shortly after the sale of the property in 1780.

THE LATE MEDIEVAL HALL HOUSE

A surprising amount of the original late medieval house survives in the

The medieval oak screen

The muntins showing mason's marks

The extended crossbeam

entrance hall, kitchen/living room and bakehouse section of the house. The major divisions of the present house, the two thick full-height cob crosswalls, are original. They define a 2-bay hall with a passage along the eastern side (the kitchen/living room and entrance hall) and single-bay inner room (the bakehouse). This probably dates from the period circa 1500-1540. The original oak plank-and-muntin screen remains in its medieval position, between the passage and hall. The muntins (upright studs) are chamfered on both sides and the chamfers are returned across the top of the plank panels on the underside of the headbeam with mason's mitres. Some of the carpenter's assembly marks can still be seen on the passage side, semi-circular scratches as Roman numerals and numbering both planks and muntins in order from the rear to the front. The original doorway is towards the front of centre. It shows evidence for some form of arched head, either shoulder-headed or low Tudor arch, but this has been cut back to a square head. It was a low partition screen, never rising above the headbeam and both passage and hall were originally open to the roof.

The major oak timbers of the medieval roof still survive. It was two bays with a central roof truss. It is a short cruck typical of the North Devon area in the fifteenth and sixteenth centuries; that is to say, the main principals curve in cruck-like fashion to sit in the top layer of the cob side walls. The collar is cambered and it carries two (probably three originally) sets of threaded purlins and a threaded ridge. The lengths of purlin are scarfed together as they pass through the truss principals whereas the lengths of ridge are mortise-and-tenoned together. These timbers are heavily sooted because the original fire in the hall was an open hearth set somewhere in the middle of the hall floor; its smoke simply rose into the roofspace and escaped through the thatch.

The hall was the main room of the medieval house. Food was cooked and eaten here and servants probably slept here too. The usual arrangement was for a bench at the upper end (the crosswall opposite the passage) with a table in front.

The inner room end was two storeys from the beginning, providing a storage room below a first floor bedchamber (sometimes called the solar) for the master of the household and his immediate family. The doorway from the hall to the inner room still contains its original oak frame - with a particularly well-preserved Tudor arch. The plain chamfered beam over the inner room and some of the roof timbers (which are clean here) are probably late medieval but no evidence shows for the position of the stair or ladder to the bedchamber.

The massive side oven in the bakehouse

At the other end the full height crosswall on the lower side of the passage is thick enough to have been an external wall, and a straight join can be seen in the back wall. It may well have been the end of the domestic section of the house but it is usual for there to be another room below the passage in the late medieval period. This room was usually of low status; either service, storage or even agricultural in use. The construction of the late eighteenth century parlour has removed all evidence for such a third room.

SIXTEENTH AND SEVENTEENTH CENTURY IMPROVEMENTS

Among the first of a series of early modernisations was the insertion of a chimney stack to provide an enclosed fireplace for the hall. This was a lateral stack projecting from the front wall alongside the front doorway; some of its masonry still shows in the front wall. This fireplace was demolished in the late eighteenth century and all that remains is the right side and back of the fireplace plastered over in the front corner of the hall. There is nothing physical to go on in terms of dating this fireplace but the second half of the sixteenth century is the usual date for the replacement of the open hearth by an enclosed fireplace.

The hall apparently remained open to the roof at this time although a small first floor chamber was built over the passage. Only one joist remains notched into the top of the screen headbeam. It is chamfered with step stops typical of the sixteenth century. The series of notches along the top of the headbeam might indicate that the passage chamber oversailed the passage screen and jettied into the lower end of the hall.

Some time later, probably in the early-mid seventeenth century, the hall was floored over to provide an extra bedchamber. The crossbeam in the hall is chamfered with pyramid stops and dates from this period. However it was built into the breast of the front fireplace and, when this was demolished in the late eighteenth century, it was cleverly extended to reach the rebuilt front wall. There is another crossbeam of about the same date against the passage partition but this appears to have been pushed in later, probably in the late eighteenth century.

The next major alteration was the Georgian refurbishment but there are a couple of other features which deserve attention. The two doorways from the former hall/present dining room to the rear outshut contain ancient, solid, door frames. The right one once had a chamfered surround and could be seventeenth century and suggests that there was an outshut there by that date. The other has a bead-moulded surround and is probably early-mid eighteenth century. There is also a good elm bench along the front wall of the hall/dining room. It is certainly old but cannot really be dated. It is tempting to propose that this was the old bench against the upper end of the hall which was moved round to its present position when the sixteenth century fireplace in the front wall was demolished and the late eighteenth century fireplace inserted into the crosswall.

SUMMARY

Devon is remarkable for the large number of similar late medieval farmhouses but each is quite different with its own special character, the product of generations of individual farming families. Huxtable was hand-built by local craftsmen using local materials and has evolved over five centuries from a medieval open hall house to the present farmhouse. Its development reflects improving standards of comfort and privacy over the years as well as the fortunes and personalities of its successive occupants.

John R.L. Thorp

Appendix 9 gives suggested plans of the Farmhouse at different periods.

Chapter 4. THE FORTESCUE FAMILY

The Fortescue family have lived in Devon and Cornwall for many centuries, but Castle Hill, the Mansion at Filleigh, the main seat, has been in the family since the marriage of Martin Fortescue in 1454 to Miss Denzell, the heiress of Filleigh and Weare Gifford. The Fortescues actually came to England with William The Conqueror. The many branches of the family have owned properties all over the West Country ever since. They first came to prominence during Henry V's French campaigns. The next generation included Sir Henry Fortescue, Lord Chief Justice of Ireland and Sir John Fortescue, Lord Chief Justice and Lord Chancellor in the reign of Henry VI. It was the latter who actually wrote the laws of England: *Da Laudibus Laguna Anglica*. Sir John Fortescue (courtesy title Viscount Ebrington) lived and died and is buried at Ebrington in Gloucestershire, which traditionally has been the home of the eldest sons until they succeeded to Castle Hill. It was Sir John's second son, Martin, who married the heiress as mentioned above. The family also had properties in County Waterford, Ireland, in Buckinghamshire and at Tattershall Castle with 500 acres in Lincolnshire too, but gradually all these properties were sold, as land around Castle Hill and Exmoor was bought whenever it came on the market. In fact, when my Father died in 1958, the whole property was nearly 30,000 acres, but sadly, about half had to be sold including Huxtable Farm and other farms on the northern part of the Estate to pay large duties.

My ancestors served Devon in all forms of public life such as Sheriff and Lord Lieutenant and many were members of Parliament and very active in the political world, and others were soldiers and sailors in the Army and Navy. Nearly all the Fortescue male heirs were named Hugh. My only brother, Hugh Peter Viscount Ebrington was killed at El Alamain in 1942 and was the last of the male line to live at Castle Hill. As the property was not entailed, my Father decided to leave Castle Hill and the North Devon and Exmoor and Challacombe Estates (nearly 30,000 acres) to his elder daughter, and to my younger sister he left the Weare Gifford Estate near Bideford. My Father gave Ebrington Manor, the land in Gloucester, to my Uncle Denzil, his younger brother, immediately after the war. He succeeded my Father as 6th Earl in 1958 and lived there until he died in 1977 when he was succeeded by his eldest son, Richard as 7th Earl. Richard's eldest son, Charlie, Viscount Ebrington, now lives at Ebrington with his family.

A manor house had existed at Filleigh for a long time and was always the heart of the Estate. It was altered several times over the years and plans show that it was rebuilt by Arthur Fortescue in 1684. However, it was his grandson, Hugh Fortescue, who inherited the barony of Clinton through his

Castle Hill before the fire of 1934

Mother, who built Castle Hill as it then became known between 1729 and 1740 in the Palladian style, and laid out the landscape. In 1749, he was created Earl of Clinton and made Baron Fortescue of Castle Hill with special remainder to the Barony to his half brother, Matthew. Hugh died unmarried in 1752 and was succeeded as planned as 2nd Baron Fortescue by his brother, Matthew. The Earldom of Clinton became extinct and the Clinton barony went into abeyance. Matthew and his wife, Anne, do not seem to have made any alterations to the mansion, but he built the Holwell temple (which is now a ruin but is going to be rebuilt by the Landmark Trust) in memory of his half brother and also made alterations to the landscape. Matthew was succeeded by his son, Hugh, as 3rd Baron Fortescue in 1785 and who was created an Earl in 1789. Sir John Soane, the famous architect, was employed to do more alterations to the mansion between 1798 and 1802. The 2nd Earl, whose long political career in the Commons terminated in the Lord Lieutenancy of Ireland 1839 - 1841 (and the Order of St Patrick and the Order of the Garter in 1856), succeeded his Father in 1841. He put in hand a second enlargement of the mansion under the auspices of the Architect, Edward Blore - the roof was raised (literally) and another floor fitted into the attics and a new entrance hall and staircase constructed as well as the addition of a porte cochere. The 2nd Earl was a considerable philanthropist, and it was he who supported and encouraged the Rev. Brereton in the founding of West Buckland School. He had a distinguished Parliamentary career as mentioned by Mrs Cameron in her chapter on the school.

Hugh, 3rd Earl, succeeded in 1861 and like his predecessors, had an important political career and continued the good works of his Father in the County and on the family Estate. He, like them, who had died respectively at the ages of 87 and 78, lived to the ripe old age of 87 and was succeeded by my Grandfather, Hugh 4th Earl in 1905. My Grandparents lived at a charming rented house called "Bydown" near Swimbridge (now turned into flats) and my Grandfather rode to Castle Hill to superintend the daily running of the Estate during the years of my Great Grandfather's life. My Grandfather again was involved in political life as a MP before he succeeded to the title. He was Lord Lieutenant of Devon and was deeply involved in all matters in the County, especially the T.A. as well as being Master of the Devon and Somerset Staghounds for many years. My Father succeeded him as Hugh 5th Earl in 1932. Again, he became Lord Lieutenant of Devon and also had a very distinguished career of military, public and political duties culminating in being made a Knight of the Garter and being one of the four Knights to hold the Canopy over the Queen at her Coronation.

A ghastly tragedy overtook us when Castle Hill was burnt to the ground in March 1934. The whole family were away at the time as

Castle Hill today

considerable alterations were being made to the mansion, but our beloved housekeeper, Mrs Vincent, and a young housemaid, Joyce Davey, who slept on the top floor (where the nurseries were also situated) were unable to escape and were overcome by the smoke and killed. Straight away, my parents decided that Castle Hill must be rebuilt and engaged the notable Architects, Lord Gerald Wellesley and Mr Trenwith Wills to draw up the plans. It was considered best to put the mansion back to its original Palladian proportions - in other words, to do without the extra floor put in by Edward Blore, and to make various other internal 20th century improvements as well as moving the front door to a more central position with a semi circular porch. (I have actually replaced this with a colonnade designed by the well known Architects Raymond Erith and Quinlan Terry). We lived at Simonsbath House (since sold and now an hotel) during the rebuilding of Castle Hill and thankfully moved back home in May 1936.

Not a great deal of work was done on the Estate during the war years as most of the staff were enlisted into the forces. Two landgirls worked at Home Farm and others for our tenants (like Mrs Ridd). German prisoners of war worked in the Woods Department and made some good roads which have proved very useful for extracting timber. Everybody in the villages had evacuee children billeted upon them, and at Castle Hill, not only did we have evacuees in the Garden Cottages, but we also had an entire Preparatory School, St Peter's Seaford, of 70 small boys in the mansion. They were an enormous asset to the village - the boys sang in the choir; the Headmaster, who was a Lay Reader, took Services at St Paul's, Filleigh; some of the Masters played for Filleigh Cricket Club, others lodged in the village; the boys grew vegetables in the formal flowerbeds in Castle Hill garden ("Dig for Victory"), and when my Father was on leave from the Army and wanted to go shooting, the boys used to "beat" for him. I quite often meet or get letters or telephone calls from now quite old boys asking if they may bring their wives and sons to Castle Hill to show where they were educated during the war.

As I wrote earlier, my Father died in 1958 at the age of 70, 4 days after the death of my Mother aged 65. They had both worked untiringly in the County and nationally in all aspects of public life as well as taking the closest interest in the running of the Estate and the welfare of all the tenants and employees. My family and I (my 2 stepsons and my 2 daughters) lived very happily at Castle Hill from 1958 until I gave it to my daughter 2 years ago. During this time, as I mentioned earlier, roughly half of the Estate had to be sold to meet the death duties payable on my Father's death. We took several farms in hand and farmed them ourselves, but there are still 14 farming tenants. Huxtable was one of the farms that was sold to the then sitting tenants, Mr & Mrs Ridd, and it is lovely to see just how successful the

Payne family have been with their excellently designed conversion.

I am glad to say that my daughter, who is now The Countess of Arran and to whom I gave Castle Hill mansion and the Estate (or what remains of it) a few years ago, is following in her ancestors' footsteps as an excellent land-owner and is very active in public life, including being High Sheriff of Devon in two years time.

Chapter 5. WEST BUCKLAND SCHOOL

The origins of West Buckland School are in some respects unique, reflecting significant aspects of mid-Victorian educational thinking. There is an intriguing connection between Dr Thomas Arnold of Rugby School, father of poet Matthew Arnold, and the three pupils who arrived in the wildness of West Buckland on a dreary November day in 1858.

Dr. Arnold had perceived that there was a need to extend the educational advantages enjoyed by the more wealthy.[1] A pupil of Arnold's, the Rev Joseph Brereton from Norfolk, succeeded in putting this philosophy into practice, though in a more restricted way than he had originally intended.

In mid-Victorian England there had been an expansion in the provision of elementary education by the government and in the creation of new "public schools". Brereton believed that these developments emphasised class divisions within British society and a large section of society benefited from neither. He identified the "middle classes" as being both rural and urban, and from professional, trading and manufacturing occupations.[2] His vision was that in the large counties of England a number of public schools for the middle classes would be established under the direction of the Lord Lieutenant and the county authorities. There would be county colleges and a county examinations[3] and he envisaged the possibility of university education for the sons of traders and farmers. By 1880, he was associated with many county schools throughout the country.

That Brereton's scheme in general did not flourish as he had hoped was the result of complex local and national factors. The locations of the county schools were remote, for instance, and the expectation that farmers would both want to and be able to support these schools had perhaps been misjudged.[2] Agricultural depression and rural depopulation also had an effect. Brereton, like Arnold, was unsectarian and therefore in a sense non-controversial, and this was praised by a Schools Inquiry Commission in 1880.[4] Yet this position might not have appealed to parents of strongly high or low church backgrounds. There was strong competition from other similar schools and from newly-founded state-aided schools.[5]

Brereton's association with North Devon began when the living of West Buckland was presented to him by Lady Bassett of Tehidy.[2] For his ideas to assume reality he needed the support and resources of persons of eminence[1] and by far the greatest and most persistent contribution was that of the Fortescue family of Castle Hill, in the neighbouring parish of Filleigh. Hugh, the Second Earl, was Lord Lieutenant of Devon and had a deep interest

West Buckland School

in education. His son, Viscount Ebrington, had been private secretary to Lord Melbourne, the Prime Minister. He was Member of Parliament for Plymouth between 1841 and 1852, held the post of Lord of the Treasury from 1846-47 and was Secretary of the Poor Law Board from 1847-51. In 1858, Brereton and Earl Fortescue published a pamphlet, *The Devon County School: its objects, costs and studies*. The school was to be an independent, self-supporting, non-diocesan foundation, its financial basis a combination of commercial and endowment principles.[5] Brereton gave a more detailed exposition of his ideas in *County Education* (1874).

The first three pupils of West Buckland Farm and County School, as it was called at first, stayed only a few weeks at Miller's farm, Stoodleigh, using the parlour, hall and a bedroom. Their teacher was J.H. Thompson, whose connections with the school, as teacher and Headmaster, continued until 1888. The fees were seven shillings a week, with extra for books and washing, but the pupils were allowed to earn their board by agricultural work. It may have been this that provoked comments from Hugh Gawthrop in the *North Devon Journal* letters columns. Gawthrop, head of Mount View Academy in Barnstaple, referred to "Juvenile slavery" and a "Semi-Union Establishment", likening it to a workhouse.

The second term opened in larger premises at Tideport with ten boarders and one day-boy.[2] In 1860 the Devon County School Association was founded to raise £7,500 through three hundred shares, of which twenty-two were held by Earl Fortescue and Viscount Ebrington. The land was rented to the school by the Viscount, with the opportunity to buy the site within five years.[1] In 1860 the school moved to its present site, where the Second Earl had laid the foundation stone. The inscription expresses *the humble hope that the Great Architect of the Universe, the Maker of Heaven and Earth, the Giver of all Good, will bless and prosper the work this day commenced...*[1]

From its early years, the school attracted much publicity and many distinguished visitors who frequently stayed at Castle Hill. Lords Lieutenant, Bishops and University Chancellors gave approval to ideas that Brereton himself acknowledged might be considered audacious and impertinent.[4] The Prince of Wales referred to a West Buckland boy who obtained the first agricultural prize in the country[4]. J.H.Thompson recalled a Prize Day when sixteen reporters had been at the school and in Barnstaple a special staff of telegraph operators were sent down so that the speeches might be reported in the London papers the next morning. On different occasions, he welcomed to the school four Archbishops and two Chancellors of the Exchequer.[1] Thomas Hughes M.P., author of *Tom Brown's School Days*, spoke at the 1868 Prize Day.[1]

Another aspect of the school that deserves attention is its close connections with the Oxford and Cambridge Local Examinations. Brereton and Earl Fortescue were among their main promoters, and even more closely involved was Sir Thomas Acland, who had also played a part in the establishment of the school. Acland, a lifelong friend of Gladstone's, appealed to Oxford in 1858 to sanction the examinations in "middle class" schools, an idea that had been put forward in earlier schemes. In 1862 West Buckland became the first school in the country to become a centre for the Cambridge examinations. "West Buckland" appeared at the end of a list of national centres, mainly large towns, and the writer of a letter to *The Times* asked, "Where is West Buckland?" R. Pearce Chope, a distinguished Old Boy who writes of this some years later, suggested that this was "a query which would even now be somewhat difficult to answer with any great degree of accuracy". However, despite its remoteness, on several occasions more boys from the school passed the Oxford and Cambridge examinations than any other school in the country.[1]

In recent years there is one name associated with West Buckland that is almost certainly more widely recognised than any other. R.F. Delderfield, playwright, journalist and novelist, is perhaps best known for *To Serve Them All My Days*, a novel which became a television series bearing the same title. By his own admission a "professional romantic", he portrays the school sympathetically through the eyes of David Powlett-Jones, teacher and Headmaster. It is fiction, yet it has been said that it is an uncannily accurate portrait of West Buckland between the last years of the First World War and the early 1940s.[6] Delderfield said that of the six schools he attended there were five that he was glad to leave, but there was one that he always regarded with great affection.[7] He remembered it as a series of discoveries, some of them magical, a few frightening, but all of them absorbing to someone with a predisposition to translate observation and experience into word pictures. There were the "bluebells covering the steeply-angled slopes", "the pleasant sound of ball snick-snacking on willow", "the massed pit-pat-pat of rubber shoes on the surface of Exmoor farm-tracks" and "a few microscopic triumphs such as running second in the Senior steeplechase".[8] There was a sense of belonging, of knowing where you stood, that made pleasantly tolerable even the less savoury parts - the smells of "cabbage water in the long funnel of a passage leading from quad to dining-hall, blanco in the old covered playground where seniors smoked ivory-tipped de Reszke cigarettes...".[9] An abiding pleasure, bringing relief to Sunday gloom, were hot pasties, apple turnovers and currant cakes sold from her farm kitchen window by Mrs Stanbury after church.[10]

Now, more than sixty years after Delderfield's schooldays, West Buckland prospers in an unchanged rural setting. The original boarding school is at the centre of an expanding complex of sportsfields and new buildings: the three hundred day pupils, whose ages range from five to eighteen, have their lunch with the boarding community in the original Karslake Hall. Of the changes, one of the major ones is that there are two boarding houses for girls. But the school day, with its particular blend of study and activities, music, debating and drama, remains largely as depicted by Delderfield.

NOTES

1. *West Buckland School, 1858-1958.*

2. Jon Edmunds: *A History of West Buckland School.*

3. *West Buckland Year Book and Kalendar for 1860.*

4. *West Buckland School Register.*

5. J.R. de S. Honey: *Tom Brown's Universe.*

6. *North Devon Journal Herald* (28th February 1983)

7. R.F. Delderfield: *To Serve Them All My Days* (foreword).

8. R.F Delderfield: *Overture for Beginners.*

9. R.F. Delderfield: *Tales out of School.*

10. R.F. Delderfield: *For My Own Amusement.*

Chapter 6. THE RIDDS AT HUXTABLE: pre 1940-1952

I am pleased to have the opportunity to write about Huxtable, which has meant so much to me. I shall attempt to create for the reader a picture of life on the farm from around 1940-1980: the acquisition of the farm in 1940; my arrival as Doreen Cass, a land girl in 1943 and my marriage to Fred Ridd in 1944; the development and subsequent efficiency of the land and stock, and the bond that formed between me and the piece of land known as Huxtable.

The Slader family were tenants at Huxtable up to 1940. One cannot help feeling a degree of sadness, that one hundred years of the Slader occupation should end with an elderly childless couple living like peasants. Poverty was evident everywhere. Granfer Ridd visited them prior to moving in. He found Johnny Slader in bed, in his dirty working shirt. He had obviously been taken ill whilst working. A bowl of unpalatable soup had been left to get cold and the room was damp and dark. There was never much heating in the house; Mrs Slader believed a fire was made of one small stick resting on another. The old man was never able to dry his clothes as he hovered over it. Constant damp clothes and malnutrition contributed to his feeble condition.

The late Jack Williams, the local butcher, could tell many interesting tales of the Sladers of Huxtable. A pig of questionable age, grovelling for a tasty morsel was on the hit list and, having been ill-fed, this resulted in the hairs sticking up on its back like porcupine quills. Jack had to deal with it as best he could. Calling in for his money, he saw the table laid for dinner. He never forgot seeing the boiled head of a sheep, with the eyes giving a far-away glazed look, staring with disdain at the dismal scene. The sheep had been kicked by a horse six weeks before and the head was probably the last vestige of a meal. Jack took his money and left hastily. One regrets the passing of such country characters as butcher, Jack Williams; fond memories of such people will long remain.

Tales from some of the workmen who helped at harvest time come to mind. Harvesting in the pre-mechanised age was a real family venture. There was nothing like meals enjoyed in the harvest field after hours of using the scythe. It was said that Mrs Slader sent out a jug of tea and rock cakes, "rock" being the operative word. Totally inedible, they were kicked around the field. Lunch consisted of the front leg of a rabbit and two potatoes. Such was the way in which they eked out a living at Huxtable around the 1930's.

Johnny Slader's days at Huxtable were numbered and he was

eventually evicted by the War Agricultural Committee in 1940.

1941-1952

The Ridd family moved into Huxtable on Lady Day, 25th March 1941. They inherited a legacy of neglect and decay on the land, among the farm buildings and in the house. Fortescue Estates settled the rent at seventy-five pounds per annum. The neglected lane leading to the farm caused problems on the first day. It was a rugged track with trees overhanging and a horse and cart was necessary to transport the furniture from the lorry to the farmhouse. The lane has always played a leading part in the life of the farm - measuring half a mile, it could tell its own story.

Chickens enjoyed the luxury of the back kitchen, and approximately twelve barrow loads of droppings were removed. The aroma of ammonia readily comes to one's nose! Lime and straw carpeted the main kitchen and bedrooms. The front garden had a display of large holly trees, preventing light and sun penetrating the south facing rooms.

It must have been a difficult time for the family (Elizabeth and Fred, Mary, Fred junior and Hilda), settling into new surroundings, making provisional plans for production and income. The workmen from Fortescue Estates moved in at the same time, to repair and replace windows and door frames, instal racks and mangers in the buildings, and erect partitions in the shippen for the milking cows. The house needed new doors and windows, and were duly painted the statutory dark red like all Estate properties. The original slate floor in the kitchen was left but the front room floor needed replacing. There were rat holes around the room, proudly displaying jagged pieces of glass, to keep out the offending visitors.

It was a lean time for the hay and corn harvest. Many of the fields were permanent pasture and would readily respond to several dressings of fertiliser. Big Field was eleven and a half acres and was thick with stroil (grass with long creeping roots). With his faithful shire horse, Darling, Fred ploughed each acre four times, one furrow at a time, followed by raking and burning the roots. This operation took several months but it reaped its own reward. As the years progressed, vegetation and crops grew in profusion, sustaining the increasing flock of sheep, herd of milking cows and bullocks. The steep field to the west of the farm consisted of dense ferns almost five feet high. The only way to clear them on such a steep gradient was by hand. Granfer Ridd and Fred were not deterred by this mammoth task. The permanent meadow grass soon responded to light and air and grazing animals were responsible for turning grass into dung, spreading it around and thus

Top: The thatched farmhouse   Bottom: Fred Ridd with his Shire horses

improving fertility.

It was down in this valley that a hydraulic ram was installed early in 1940. The "ram" was an ingenious piece of machinery. A stream was piped through, enabling the piston to pump spring water up to the corner of Barn Field, into a tank of 1000 gallon capacity. By gravity it was piped to troughs in the fields, into the house and buildings. This same system also supplied the water troughs of neighbouring farms. From time to time the stream would silt up, causing the whole operation to grind to a halt. The old hand pump in the small kitchen had to be encouraged to yield its precious and reliable supply of icy water. Granny Ridd always said that this was the best water for washing butter milk from the butter and cleaning the guts of the pig for sausage skins.

The war progressed and this brought many changes to village life, least of all the invasion of members of the Women's Land Army. What a challenge for the young lads: an injection of new blood! This is where I come in on the scene. The urban life in the suburbs of London did not really fulfil me. My decision to join the Women's Land Army changed the whole course of my life. As I alighted from the steam train at Filleigh Station in June 1943, I stepped back in time about fifteen years. My initiation into farming life began with Mr. Stanbury at Higher Pitt Farm, East Buckland, and I had found a new life that completely took me over. I became Mrs Ridd on 26th December 1944 and moved into Huxtable in January 1945.

The old house was to see another change. It was divided into two dwellings to accommodate Fred and me. With two sets of stairs it served the purpose admirably. A small kitchen was fitted, and we had an outside toilet and a washstand in the bedroom, but it was home and we were happy.

After twelve months occupying the smaller part of the house, we changed ends with Granny and Granfer Ridd. We now lived in the large kitchen with a dairy and pantry adjoining. I can recall cosy evenings with a blazing log fire in the large ingle-nook fireplace, the warm light of the oil lamp and a candle to light us to bed. It took me some time to slow my pace when opening a door, because the quick rush of air would extinguish the candle.

The winter of 1947 was a tough one. I was twenty-three and Fred thirty and we were able to tackle any job, however menial. In March 1947 Granny was recovering from a major operation and recuperating with her daughter, Hilda, in Charles parish. Granfer had injured his leg and being the tough man he was, it was difficult to keep him resting. We threatened to chain him to the leg of the chair. Temperatures were very low indeed; even the dregs in Granfer's tea cup beside his bed froze during the night. The blizzard struck with great force one afternoon; there was nothing to do except keep Granfer company around the open fire, and watch the snow completely obliterate the light in the window. As the storm abated, it was a fantastic sight to venture into a white wilderness.

The lane again comes into the picture. Lying north to south, the east winds found it a very convenient pocket to drop its white, cold commodity. Fred and I dug a pathway through the lane to the road in order to carry out the churns of milk. The roads were cleared and traffic was moving long before we could.

The stable, a fortress against the cold winds and snow, housed three shire horses, Darling, Dolly and Tommy. Through lack of exercise, Darling was taken ill with stoppage. Treatment was to no avail and she died. Fred was very upset; memories crowded in upon him of the years of faithful service she had given. We had no tractor or car. Our neighbour, Arthur Pike, helped out by dragging her out to the road to a waiting lorry. It was distressing to see our dear old Darling leaving us in this way.

The shippen adjoined the house, where we had eight cows sleeping in overnight. In the front room we could hear the rattle of chains, as the cows rubbed their necks against the stanchions. I can still smell the hay pitched down to the waiting cows, mingling with their warm breath, as they contentedly munched their feed, uttering deep moos of satisfaction.

In the autumn of 1947 we employed a thatcher to renew the front part of the roof. This is truly a countryman's skill. The work was well under way when I came home with our new daughter. Margaret was born on 17th November 1947 in South Molton Cottage Hospital. Maybe the first time for many a year that a young baby had taken up residence at Huxtable Farm.

Money was not at a premium, as the following balance sheet dated 12th February 1949 shows:

Liabilities		Assets	
Capital Account	17s 7d	Cash at bank	17s 7d
Loan - Mrs M E Ridd	£1524 8s 3d	Stock on hand	£1171 6s 9d
		Implements	£353 1s 6d
	£1525 5s10d		
			£1525 5s10d

The value of the stock and implements were part of the loan from Granny. In 1949 we replaced the horses with a tractor. It is interesting to note that in nine years the farm was beginning to support an increasing number of stock, as the following list will reveal:

Live Stock at Valuation on 12th February 1949:
Sheep - 47 ewes in lamb, 25 ewe hogs, 3 ewes, 1 ram.
Cows - 10 cows and heifers in calf and\or in milk, 3 maiden heifers, 5 heifers and 1 bull calf, 1 light weight cart mare 12 years old, 1 sow and 8 piglets.

Granny Ridd was called to rest on 21st November 1964. Our second daughter, Christine, was born at home on 7th March 1952. Our family was complete and the girls grew up to enjoy life on the farm.

Young rams in Barn field

Chapter 7. THE RIDDS AT HUXTABLE: 1952-1980

Huxtable was continuing with its new lease of life and the toil of the 1940s was showing results. The hedges were strong, the gates hung well, seed and manure was sown and the stock were well nourished. Fred and I with our two daughters were optimistic about the future.

The following receipts for 1952 were encouraging and we now had more than 17s 7d in the bank!

Milk	£311	3s	4d
Wool	£185	0s	11d
Sale of sheep	£372	5s	7d
Sale of cows	£159	0s	0d
	£1027	9s	10d

In the spring of 1955 ill health hit the family again. Fred was laid low with farmer's lung. This condition is caused by the spores from dusty hay, settling and multiplying in the lungs, causing acute breathlessness and fatigue. Unable to work for seven months, he slowly recovered but his condition caused problems throughout the rest of his life. The cows needed milking and during March seventy-five ewes were due to lamb. Fred helped with advice and encouragement but physically he was unable to do very much at all. One ewe having difficulty giving birth was coaxed into the kitchen where the expert hands of the shepherd gave the urgent attention required. My sister, Irene, came down from Surrey to stay for a month, to help with meals and the family, which allowed me to work full time on the farm. The elements were very severe one morning, resulting in part of the cob wall collapsing on the back doorstep. At the same time the old sow got into the small field where there were ewes and lambs and she had to be removed urgently. Sheets of rusty galvanised iron were wrenched from their moorings, as the wind screamed and whined - quite a day to remember!

In spite of ill-health we never gave in and records show that in 1959 we had the following stock: 120 ewes, 117 lambs, 8 cows, 3 in-calf heifers, 15 cattle, 1 horse, 80 hens - total value £2354. In 1960 the rent was £215 per annum.

It was in the mid-1950's that we were made aware of the Hill Farm and Livestock Rearing Act 1946-1956. This was a farm improvement scheme. All work had to be carried out in accordance with estimates, plans and specifications approved by the Ministry of Agriculture, Fisheries and Food

(MAFF). As a result of this scheme the lane was completely renewed. In the house a new Rayburn and hot water system was installed and a bathroom fitted. By September 1960 we were connected to the mains electricity. From candles to electric light - it meant guaranteed good light in the dark hours and power to drive farm machinery. In the house my top priority was a deep freeze. I craved for an end to salt pork, for which I had never really acquired a taste. The next pig was destined for an icy grave.

Additional improvements were made on the roof of the farmhouse. The house was thatched and was in a poor state of repair. The work to re-roof with concrete tiles started on September 27th 1960. We had twelve men working up there, stripping off the old thatch, which in some places was three feet thick. It was quite a shock to see the condition of the roof timbers, eaten away with dry rot and woodworm and a startling revelation to see the charred beam in the chimney at the east end of the house. The men had to endure appalling conditions and covered a day's work of removing the thatch with tarpaulins. It was a fascinating tune as rain seeped through ceilings to waiting receptacles, from china pots to tin buckets. Ceilings suffered from the weight of men erecting new timbers and all needed renewing or patching. The work was completed on 17th October 1960. Now we were able to clean up and plan new decorations. The previous policy to paint Fortescue Estate properties a dark dull red had changed to battleship grey. I nearly exploded - and have the entry in my diary to prove it! We agreed to paint the house ourselves in jasmine and white, the paint being supplied by the Estate.

Mains water was piped down to the farm in 1965 and Huxtable was registered as a dairy farm on 19th November 1965.

The new roof was giving cause for concern. The framework was resting on the outside walls, slowly pushing them outwards. Urgent work was needed and after much discussion with Fortescue Estates, the work began again. On October 11th 1962 scaffolding was again erected and the roof removed. The first framework had not been tied in correctly and some of the joints had moved four inches. New brick pillars were built on two of the main walls to support the weight of the tiles. We were horrified to see loads of old thatch left behind on the ceilings with new electrical wiring underneath. Not only did we have tiles on the lawn but new mountains of old thatch appeared and had to be burnt. This work was completed on 2nd November 1962.

In the photograph of the farmhouse, 1956, a third tall chimney can be seen on the left. This chimney served the furnace which was used for washing clothes in the days before electricity. It would appear unique that the old house boasted two brick ovens. These were set back to back in the thick walls. They each opened out into a fireplace allowing smoke to escape up the chimney during the heating process. This was done by placing a faggot (a large bundle of sticks or twigs bound together as fuel) inside the oven and setting light to it. When reduced to ash, this was scraped out, leaving hot bricks, on which loaves, cakes and pies were baked. One of the ovens was sealed up behind the wall where the Rayburn was fitted. In the open fireplace in the back kitchen the oven was a handy store. To reach the back it took the length of my arm and a broom handle. One day when cleaning it out I could just see a piece of wood right at the back. It took some time to retrieve it and, what a surprise, it was a small pistol! My mind had often gone back to that old oven, wondering who put the pistol there and why. Years later it was identified as a muff gun, 1830.

On 20th February 1967 we received notice that:

"Huxtable Farm, East Buckland, situated in the Rural District of South Molton, has been included in the list of buildings of special architectural and historical interest in that area." (Town and Country Planning Act 1962)

On a farm all the family have to take their part in the daily work. It is also exciting to allow new ideas to develop. In the 1960s we became motivated to begin a new venture, farm holidays. We charged 6 guineas (£6 6s 0d) in 1960 for a week's bed, breakfast and evening meal. The visitors' book brings back memories of happy times and sufficient recommendations to be proud of. It was a family enterprise, offering family holidays, where children were welcome to enjoy the everyday life of a farm. The girls helped when home from school/college and Fred supervised the barbecue supper we held each week on the lawn. This proved to be very popular. After the meal Fred took the visitors on an evening conducted tour over the fields and demonstrated the skill of his dogs, as they responded to his gentle command and brought the sheep to him. This diversification of farm holidays lasted for seventeen years.

No record of our time at Huxtable would be complete without reference to Fred's passion; his sheep and his dogs. The names of the dogs are worth recording: Spot, Skye, Sailor, Shadow, Sport, Panda, Rosie, Scott and the terriers, Snip, Trixie and Paddy. Fred was a keen member of the Devon Closewool Sheep Breeders' Society and the flock were all registered as a true Closewool breed, with tags in their ears. We bred our own ewe lambs to keep up the number of breeding ewes and this meant replacing the ram every three years. This task needed an expert eye to select the perfect ram and the photograph shows one of our prize rams in 1971. A group of young rams in Barn Field shows what the farm produced.

In the 1960s under the auspices of the MAFF farmers with less than 150 acres were encouraged to participate in the Small Farmers' Scheme., Through this scheme a farm business plan was agreed and on satisfactory completion of the work a grant was paid to the farmer. We took part in this scheme and one of the objectives was to increase the poultry unit to 300 laying hens.

In the earlier days I can remember taking great care to make nests for the broody hens and always admired the patience of the hen, to sit contentedly on her eggs for three weeks but she was well rewarded with a clutch of fluffy chicks. It was a great joy when a hen would steal away, lay her eggs in a secret place and proudly walk into the yard at the appropriate time with her new family.

As we enlarged the flock by purchasing and rearing day old chicks, we were helped by the National Advisory Service and kept records on a commercial basis. It was interesting to keep detailed records as the following example will show:
1st August 1963 - 8th June 1964 (315 days) - 3,408 dozen eggs sold - profit 14s 9d (75p) per bird.
9th June 1964 - 28th September 1964 - 3,060 dozen eggs sold.

We reared caponised cockerels for the Christmas market, producing a bird of eight to ten pounds and had many local orders. In the records of Christmas poultry for 1957 we charged 2/6d. (12.5p) per pound! Preparations for Christmas meant feathers amongst the Christmas cards but the end result was well prepared birds ready for the oven. Great care was taken in dressing a bird, with the liver tucked under one wing, the gizzard under the other wing and the lace effect of the fat was deftly spread over the breast but the chief skill was to skewer the legs right back in order to protrude the breast.

In 1970 the Fortescue family had to sell many of their estate farms and land to pay death duties. We were not on the list selected at that time but our request in 1975 to purchase Huxtable was accepted and, as sitting tenants, a price was agreed. We signed the contract and paid the deposit on 17th July 1975. This ended a period of ownership by the Fortescue family of nearly two hundred years.

Gradually as time and money permitted and the commercial market dictated, the farm was brought through various phases of improvement and modernisation. Milk was now a profitable business. We started to increase the dairy herd and during the summer of 1971 we made well over one thousand bales of hay. We hired a contractor to bale the hay and he charged 3.5p per bale. I think back to earlier times and how difficult it must have been to make hay in 1940 when weeds choked the grass. I refer to this was evidence of the reward for Fred's hard labour with his horse.

It was during the mid to early 1970s that Huxtable probably reached its optimum output during our time. We had a herd of 14 cows, 120 ewes, 150 lambs and 6 steers. The fields on Huxtable were small; the largest was Big Field at eleven and a half acres, the smallest, Pump Splat, at three quarters of an acre. All the fields were surrounded by well made and carefully maintained hedges, which provided valuable shelter for animals and crops. Fred would spend weeks each year re-building hedges, using traditional methods of dry stonewalling, "steeping" with turves and layering the bushes on top of the hedges. At seven hundred and fifty feet above sea level and near the Atlantic seaboard, wind and rain are common during the winter. In the days before indoor lambing, many lives were saved by protection offered by these huge sturdy banks. One cannot help wondering what he would think of today's farming methods in comparison. Each field had a very distinctive name and assumed its own character and contained a world of its own. Fox Hole Field was tilled to corn one year and I can remember hundreds of rabbits escaping as the corn was harvested. Star Park Down was open to the elements and on its north western side it was very steep, sweeping down to Orchard Piece. Next to this was a much warmer field, Panny Field. It was square and formed a pan shape in one corner. It was in this gentle field that we grazed ewes and lambs because there was plenty of shelter. A large shed in Land End Field was used in the 1950's to store sheaves of corn and each year there was plenty of activity as the threshing machine came and took days to complete the work which today is done in one go by the combine harvester. It was in Garden Field that the mangold cave was built. Mangolds are large root crops, stored in a cave made of earth and straw to protect them from the frost, until needed to feed stock in the winter. The steep field, Cow Field, adjoining the west side of the yard had an old quarry where stone was taken to help build West Buckland School. From Barn Field, the highest point of the farm, we had spectacular views of Dartmoor and Exmoor.

A feasibility study revealed that on eighty acres the farm was incapable of maintaining a further increase in stock which would be required to keep it a viable enterprise. A decision was made to reduce the dairy herd and a very slow decline in the business began. Fred's health was not a big problem but he found working in a confined area with cows and the resulting dust affected his breathing.

As I begin to conclude these two chapters, many anecdotes come to mind. I have already mentioned that the lane could tell its own story! The winters of 1947, 1962/3, 1978 and 1979 wrought havoc with the road surfaces. Our lane was one surface that did not improve by the amounts of snow we had to clear. Grants were available to farmers undertaking work of this nature. After the blizzards of January 1979 it was evident that something had to be done to ensure the future of the lane. A field officer from MAFF came to inspect the top part just after snow clearing had finished. We qualified for a 50% grant. On Monday 9th July 1979 re-surfacing operations began. Only one third of the lane needed attention, the remaining two thirds only had a few potholes to be filled. Friday 13th July 1979 was a very hot day and the men were handling tarmac of two hundred degrees centigrade. The top lane was not only a snow-trap but also a sun-trap! I can remember making comparisons between the seasons; from hot tea and rock cakes, straight from the oven this time, taken to the men standing on six foot snow drifts to this hot day when Fred took cans of cold beer up to the sweating men.

In 1980 Fred made the big decision to sell Huxtable. On 23rd July the auctioneers came to value the farm and arrange to advertise in local papers. Dr and Mrs Payne came on 2nd August to view and a deal was made on 9th August 1980.

Our clearance sale was arranged for Michaelmas Day, 29th September 1980. For days and weeks we prepared for the day. Fred was busy sorting out anything from rusty chains to the tractor and from his two year old sheep dog, Scott, to his five year old dog, Skye. What we had collected over a period of forty years was amazing. Ladders, corn bins, sheep troughs, barrels of oil, seed fiddle, two long tables from the farmhouse, bedroom furniture, bedding, surplus crockery etc. all had to be allocated lot numbers in preparation for the sale. The sheep came in for much sorting out and placing in order of age, in pens of ten each. They looked perfect specimens, revealing good shepherding.

Cardboard boxes were suddenly in vogue and I can remember collecting them from many sources. Moving day was 6th December 1980 and keeping ourselves busy with all the usual jobs helped to soften the trauma of leaving the farm. One memory will always remain with me of our last drive up the lane; neither of us spoke as we began our final journey from Huxtable. Arriving at Foxlease in Lapford, a large house and garden, the family were waiting to help and our new life began.

In these two chapters I have attempted to record the highlights of thirty-nine years at Huxtable Farm and dedicate it as a memorial to my late husband, Fred, who was called to rest on 24th August 1982. He educated me from a raw recruit from the city into a fully fledged farmer's wife. I have nothing but happy memories of those thirty-six years and I would like to thank my two daughters, Margaret and Christine, for their help in writing these chapters, reading and collating old records and photographs.

Chapter 8 THE PAYNES AT HUXTABLE

When the Ridds left Huxtable on 6th December, 1980, we, Freddie and Barbara, were unable to move in immediately. The purchase of the farm coincided with the recession of the early 1980's, interest rates were up to 16% and the prospective buyers of our house near Bideford pulled out. However, despite solicitor's advice to the contrary, a bridging loan was arranged and we decided to move near the time of our son's return from New Zealand where he was "relief" farming. We used the time prior to this to restore the main room of the farmhouse - the original hall. We had been told by Doreen Ridd and older neighbours of the existence of, at least, one large open hearth and of how the house was thatched until 1960. We also suspected the presence of a screen at the side of the cross passage. Mr Charles Hulland, an authority on Devon farmhouses, was contacted for advice and, prior to his visit, the cement blocks and a Rayburn in the main room were removed to reveal a large open hearth.

His visit was dramatic. When we bought the farm the house was immaculate and the changes we planned - a large kitchen to be made in the adjoining dairy and store room and the opening up of the fire place - had already created much dust and dirt. But Mr Hulland's visit was to reveal that there was much more to be discovered. On arrival he donned a woolly hat and slippers and disappeared into the roof space where he reported that the original smoke-stained timbers could be seen and the carpentry of the joints suggested a date of 1520. We next removed the vinyl from the floor of the main room and several layers of linoleum and newspapers, in the hope that we might find traces of the original fireplace from which smoke would have gone straight up and out through a hole in the thatch, so causing the staining on the roof timbers. This was not to be so although we were pleased to find an intact flagstone floor.

Then Mr Hulland turned his attention to the wall between the main room and the entrance. It had been suggested that we might discover a screen and this is exactly what happened, for, when Mr Hulland stuck a screwdriver into the carefully papered wall and pulled away the plaster board, a very dirty oakscreen painted red was revealed! Many gallons of Nitromors later, a screen, the length of the house, was uncovered and, in the same way, beams and benches were stripped of their coverings, cleaned and polished. There was also evidence of a smoking chamber but no opening from the main room was found for this. The chapter by John Thorp explains Huxtable and enlarges on what we learned from Mr Hulland and, later, from a Mr Mair who was an architect surveying listed buildings for Devon County Council. The work of stripping and cleaning took many hours and when we eventually

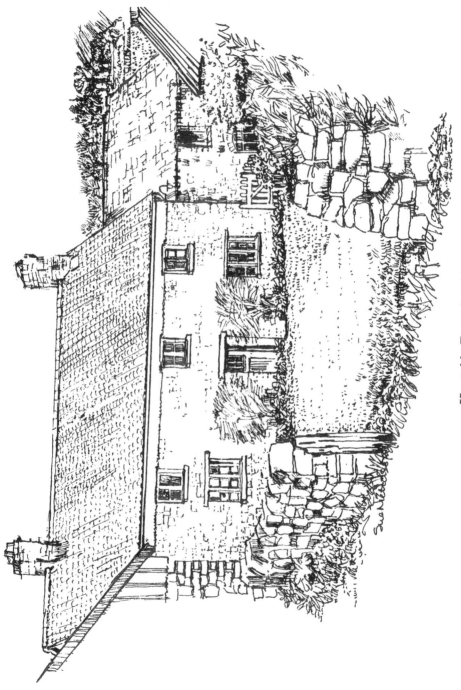

Huxtable Farm in 1980

moved into Huxtable only this one room was finished. Antony returned from New Zealand and, as we were the legal farmers, he became a very badly paid farm labourer responsible for establishing a viable farm.

In the two years before Jackie takes up the story much happened. We arrived with our only livestock, one elderly dog, one young cat, two foals and one pony, whilst our "machinery" consisted of our old Landrover, a horse trailer and a Mini. First I had to learn about P.A.Y.E. and V.A.T. and take over the farm accounts. Our youngest daughter was still living at home prior to university and we all helped on the farm in whatever way we could. Freddie and I both had demanding professions, he as trauma consultant at the North Devon District Hospital and I as counsellor in one of Barnstaple's comprehensive schools.

The weather the first winter was appalling. We were snowed in with no electricity and telephone for several days and it was then that we started to learn of the hardships involved in farming - all the first litter of piglets died one by one, sheep were trapped in the farthest fields and both Freddie and I were unable to get to our work in Barnstaple. Farming was the least profitable enterprise we had ever been engaged in - it was also the hardest work - but, as we became more and more involved in the life of West Buckland and in the re-stocking and equipping of the farm, Huxtable gave us all an increasingly rewarding life-style.

In order to help with the finances of the farm we started a Bed and Breakfast (and Dinner) business. This was during the first summer after our previous house was sold and when we were able to undertake more restoration and modernisation and plan the first conversion of a farm building. Both Freddie and I retired - he in 1982, although he continued to do locums until 1992, and I in 1984. From this time onwards the guest house side of the business continued to grow and Antony had met Jackie Devereux, a teacher at South Molton Comprehensive School, and it is she who continues with the story of how Huxtable is as you see it today.

THE YOUNGER PAYNES

It gives me, Jackie, great pleasure to be able to write about our life at Huxtable Farm. The experience of starting this new enterprise and life on the farm has been both enjoyable and rewarding, but full of surprises for us all as a family. We learned a lot the hard way, by making mistakes, and more mistakes, to accumulate the necessary experience to run an efficient enterprise and at the same time give our animals the love and care all living creatures demand. These experiences are not easily forgotten but they can be

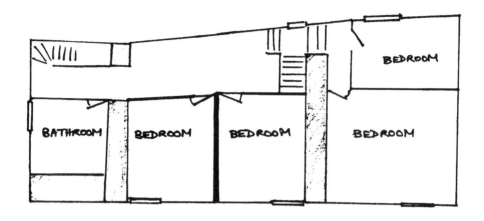

First Floor

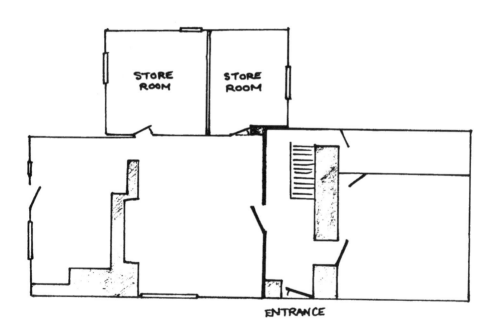

ENTRANCE

Ground Floor

Layout of the Farmhouse in 1980

52

expensive in terms of money, time and temper! This is a time when farming is under increasing pressure from countryside conservationalists, economists and animal welfare lobbies. We constantly look at our system of management of land and animal. I endeavour to illustrate in this chapter how we have built up the farm to its present day standard.

LIFE BEFORE HUXTABLE FARM

Antony's previous farming experience and interest in farming has always been evident. He was born and lived in Gambia for four years whilst his father, Freddie Payne, was in charge at the Queen Victoria Hospital and where Barbara had a private school. The Payne family always kept a few animals, including a little bambi deer, whilst in Gambia. After Gambia they moved on to Aden for nineteen months, then back to England to The Old Parsonage in Horwood near Bideford. Freddie was the Senior Casualty Officer at North Devon District Hospital and Barbara taught cookery at first until she became counsellor at Park Comprehensive School in Barnstaple. Antony lived in Horwood from eight years of age until moving to Huxtable Farm in 1981.

Living next to a farm in Horwood meant that Antony spent many hours helping out, but his main interest at this time was horses. With stables nearby, he spent a lot of his time mucking out, grooming and attending point-to-points. On his twelfth birthday he was given a 12.2 pony called Browny and later Pink Panther, a 14.2 pony. At the time he was training for the county tetrathlon where he rode Wallbrook which was eventually ridden by Sheila Carr who had just won the Junior European Championship. With this great interest in horses Antony went to Windmill Hill, Warwickshire, for two months to obtain an A.I. Certificate for Equitation, Stable Management, Horsemastership and minor ailments to qualify for teaching riding.

Before going to Seale Hayne Agricultural College in Newton Abbott, South Devon, to do an O.N.D. course for four years, Antony had to do a year's work on a farm. He did this on Henry Hazel's large farm in Holcombe Rogus near Taunton, Somerset. This is a mixed farm where Antony learned a wide variety of different types of farming - arable, sheep, milking, bullocks - and the social life of a young farmer! Whilst at Seale Hayne, Antony had a year's work experience at a very large farm, Manor Farm near Winchester.

LIFE AT HUXTABLE FARM

Antony completed his four-year O.N.D. Agricultural course at Seale Hayne College and whilst preparing for his seven month trip to New Zealand he taught at the Pony Club Camp held at Mr and Mrs Ward's, Westacott Farm, East Buckland, (now our neighbours). It was whilst there, that in July, 1980, Freddie, Barbara and Antony looked at Huxtable Farm with the view to purchasing it, as they had the opportunity of selling their family home too large now that the children had left. A deal was made and on 6th December, 1980, Barbara and Freddie bought Huxtable Farm whilst Antony was in New Zealand. Antony has many a story to tell of his time in New Zealand, milking cows for a farmers' relief service, one of them being the time he caught Leptospirosis, a potentially fatal 'flu-like disease caught from the urine of cattle.

Antony returned to England at the end of February 1981 to use his agricultural skills in running the farm. At this time the stock consisted of two foals called Rupert and Marcus, which Barbara and Freddie had purchased for Antony to break, and a well-used Landrover and horsebox. Purchasing the farm had its difficulties, for with a bridging loan, there was little money to spend on the farm. With Barbara and Freddie still working it meant that Antony was in charge of the farm and was employed by his parents to do so.

With not a tractor or any other farm machinery on site, Antony's first job was to use his childhood savings to buy the little red 165 Massey Ferguson and linkbox secondhand from Glidden and Squires in Barnstaple. Little did Antony know at the time that this tractor was to bring us many fond memories, one of which was when it was used on our wedding day. It had been given a new coat of paint and decorated with flowers for us to travel on, from St Michael's Church, East Buckland, to West Buckland Village Hall for our reception.

In July 1981 Freddie was given two young Large White gilts as a house-warming present. These were kept in the lower part of the big barn until they were bigger and moved into the pig sty. The sow became pregnant and it was during the blizzard at the beginning of 1982 that she decided to farrow. The blizzard caused an electricity power-cut and the wind made the temperature very low. Hours were spent, in vain, sitting by the Hamco oil-fueled cooker in the kitchen trying to keep the little piglets alive, followed by the mother (who subsequently died) getting mastitis because nobody could get out to the vet for penicillin in the snow-filled roads. Not the nicest of starts to stock-keeping!

For the first two years at Huxtable Farm Antony had no stock of his own. He spent those years working as a farm labourer on a farm at Brayford during lambing and silage time. To keep the grass down on the farm he let the ground for grass keep to local farmers. One farmer kept bullocks in the

Cleave, and some of them got bracken poisoning from eating the bracken during the drought summer of 1982 when grass was in short supply.

In 1980, the eighty acres of land were not fenced and had small gateways. When the Paynes arrived it had some well-maintained hedges which Mr Ridd, the previous farmer of Huxtable Farm, had obviously spent many an hour banking, steeping and laying. Mr Ridd's Closewool sheep with their docile temperament were kept in the fields by the hedges. However, the large Welsh Half-breeds that belonged to one of the farmers who took grass keep, with their breaking temperament, used the hedges for shelter and for climbing through, and not as field boundaries. Eventually they roamed the majority of the farm and they even ventured into the Fortescue woodland and neighbouring farmland! This meant that when Antony purchased his first sheep, many an hour was spent chasing stock back into their fields and tying wire into the hedges only to find that the sheep had found another hole to escape through!

In 1980, although the acreage was the same as it is now, there were more fields than there are today. Some of the hedges were in a very poor state after the Welsh Half-breeds had used them, so one of the first things Antony did, in the summer of 1982, was to remove a couple of hedges, to produce what is now known as Big Field. Antony then employed some school boys from West Buckland School, who had finished their exams, to pick up stones from the field, paying them with food and drink!

In the autumn of 1981 Antony joined South Milton Rugby Club where many of the players were local farmers.

At about the same time, dear Alpha, a Jersey housecow, was bought from Dean Farm near Parracombe. She was a beautiful heifer with a black tint to her golden brown face. She was kept in the paddock in front of the Old Shippen. Twice she failed to fall into calf with artificial insemination. She wanted the real thing and broke into the neighbours' field where she had a fling with the large cream Charolais bull. It was quite a story trying to get her back! Ross, a young girl with little farming background knowledge, had just arrived to help with the Bed and Breakfast and was promptly taken into the field by Antony to help retrieve Alpha. Predictably she was in rather a state of shock after the event, realising that she had been tapping an enormous bull on the nose whilst Antony dragged Alpha away.

Alpha by now was in calf and calved a handsome bullock, named Beta, during the early hours of 15th March, 1983, just before lambing. Beta stayed with his mother for the first two days of his life whilst she cleaned him and he drank her rich colostrum milk. He was then taken away from Alpha to a lively warm freshly-strawed cubicle where he received lots of loving attention and was reared on Alpha's rich milk. Alpha, of course, was sad to lose her calf but mooed contentedly whilst being relieved of her enormous quantity of milk. She was milked by a single milking machine in the Old Shippen, producing four gallons of milk each milking twice a day, so was always ready to come in! Beta had one bucket of milk and the remainder was either drunk fresh or made into clotted cream and put on the side-table for the guests. The skimmed milk was given to the two weaners kept in the pig sty (Byre) which had been bought to fatten on Alpha's plentiful milk and scraps left over from the "Bed and Breakfast" dinners. Beta was reared on Alpha's milk for four months until weaned and put out to the grass. He spent another two years with us on the farm before going to Taunton Market in July, 1985. It was very sad to see him being loaded into the haulage lorry.

Jack Williams, the same butcher that Mr and Mrs Ridd and the Sladers had used, butchered the pigs fattened on Alpha's milk at the end of the summer, along with any lambs that we wanted putting in the freezer. It was always a pleasure to visit Jack. He was a true Devonian with many a story to tell about Huxtable Farm and its neighbours. I remember visiting him one summer to find him up on a Devon bank with a scythe cutting the grass for hay. "I'm not up 'ere 'cause there's not enough 'ay in fields," he explained. "There's lots o' lovely herbs an' flowers in the bank, the cows love it and it keeps 'em healthy." He was a friendly, knowledgeable man, of whom we all have fond memories.

The sheep that were about to lamb when Alpha calved were an old flock bought from Blackmoor Gate Sheep Sale in the February of 1982. Antony used the money earned from working as a farm labourer to buy thirty couples, i.e. ewes with lambs. From this flock twelve ewe lambs were from the 1983 lambing and those of the original old ewes that still survived were kept for the following year's lambing. A bottle fed tame ram lamb was also kept, named Bassy, and given to Jenny, Antony's sister, as a thank-you for her assistance during lambing. He was vasectomised and kept on the farm as a teaser ram, i.e. he used to go in with all the ewes to get them excited, all cycling together, just before tupping time. He was always quite exhausted by the time the rams went in with the ewes in November. Bassey stayed on the farm until a ripe old age of eight years, when he decided to go 'walkabouts' and tease a few of the neighbours ewes for the remainder of his life!

On my arrival to North Devon and being new blood in the area, I was a challenge to the local lads, probably having the same experiences as Doreen Ridd had on her arrival to North Devon. It was in the Autumn of 1982 that

Antony and I met. Like Doreen Ridd, my youth was spent in the suburbs of London until I moved in 1978 to do a four year Bachelor of Education Honours Degree at St Luke's College, Exeter in Devon. Having experienced life in Devon whilst at college, I was reluctant to return to London and applied for an interview in South Molton. I was appointed to a teaching position in South Molton Comprehensive School and Community College, teaching Mathematics and General Science.

I shall never forget my first journey down Huxtable Farm lane, it was a dark, wet night and the branches hung into the lane hitting the car as we drove along. With the speed we were travelling I held on tight to my seat, hoping nothing was coming in the other direction and that the little rabbits that jumped out of the banks would escape the car wheels. The lane was narrower then than it is now because the large lorries, tractors and trucks that have used the lane since have taken the lower part of the bank away! This has not stopped the wild strawberries, primroses, violets and other wild flowers growing in the banks, or the blackbirds from nesting and other wild fowl and small mammals making their homes in its hedge and banks.

The lane was part of the big field in those days, with a gate at the farmhouse end of the field, so, after racing down the bumpy lane, avoiding rabbits and sheep I had to get out of the warm car into the cold, wet, windy, black night and try to open this gate with only the head lights of the car to guide me through the puddles and find the gate latch! Fortunately, Antony has fenced the lane since then and removed the gate making the entrance and exit to the farm a lot more civilised!

During my first visit to Huxtable Antony gave me a guided tour of the farm accompanied by Ben, a very old sheep dog who was blind in one eye, on loan from a friend to help round up the stock. None of the fields had stock fencing, as they have today, just the lovely banks and hedges you can still see behind the fences. As we walked round Antony gave each field a distinctive name. The 'Cleave' with its bracken covered slope leading down to the stream and wooded corner has always been our favourite field with its charm and character. Antony explained that, to be profitable, this field would require labour intensive work clearing the bracken by hand. The 'Cubicle Shed' field, now known as 'Silage Pit' field because of the silage pit dug into it next to the cubicle shed in 1986, was badly covered with docks. 'Barn' field, named because of the large barn, had then and still has at its highest point, near the water tank which still gravity feeds the stone troughs in all the fields, the most wonderful panoramic views of Dartmoor and Exmoor. The presence of the hedges which have been removed to produce 'Big' field could still be seen on the ground where the grass had not grown. 'Middle' and 'Corner' fields had

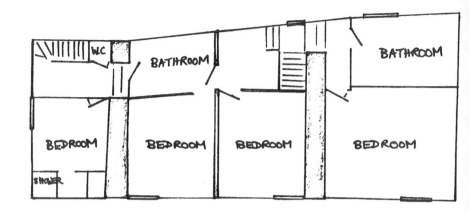

First Floor

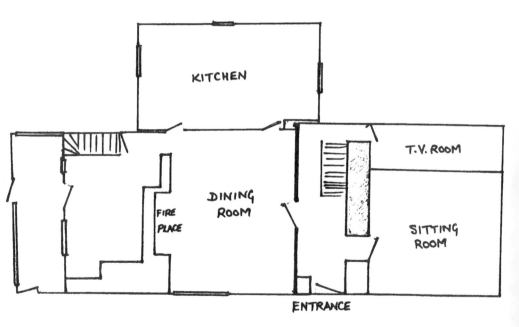

Ground Floor

Layout of the Farmhouse in 1990

a good covering of grass and as we walked along, Antony seemed to be forever bending down to pick up stones and throw them in the hedge so as not to damage machinery used on the fields in the summer months. 'Hedge' and 'Far' fields were stocked with sheep and look very much the same today as they did then, although the hedge in the middle of hedge field look in better condition then!

Back in the farm yard all the stone buildings around the farmhouse garden were agricultural buildings. There was a milking parlour, the old shippen, with a hay loft next to the farmhouse, (now home for ourselves and our four children). This was all ready to be used when Alpha calved in the spring. The first floor of the big barn, opposite the farmhouse, was half full with hay ready for the winter and the round house behind was Alpha's shelter (now Barbara and Freddie's home). The three little rooms on the ground floor of the big barn were used as a toolshed, a dog's sleeping quarters by Ben and a store for paints and other decorating materials (the present family rooms for guests). The building which is now the store and log shed was a stable for the two foals, Rupert and Marcus, with the hay loft above. There was a pig sty, now a family room for guests. The cubicle shed, the same today as it was them, was being prepared for housing the sheep during the winter months and the present day tool shed was an open fronted barn used mainly as a garage for the Landrover and tractor.

In the spring of 1983, before milk quotas were introduced, advice was taken from ADAS and accountants with plans to start dairying, with the idea of modifying the cubicle shed and to purchase 50 friesian heifer calves, rearing them to calve at two to two and a half years old. During this time we were to build the parlour and silage clamp so that milking could start when the heifers calve down. Starting dairying by this method would have meant that there was no farm income for two and a half to three years. The plan would have depended entirely on income from "off the farm" activities. An estimated cash flow was calculated for buying and rearing the heifers of at least £15,500.00 for the three years, not including cost of buildings and other expenses such as mortgage, machinery costs, contractors charges and living expenses. Although a milk quota was applied for, with this financial advice, it was decided to concentrate on sheep and to simplify the system in order to release Antony for as much contract work as possible. This was probably the best decision made because the farm does not qualify for hill subsidy as do neighbouring farms and the ground is not rich, therefore requiring a lot of fertilising and re-grassing.

During our courting years, 1983 and 1984, Antony continued contracting himself as a farm labourer, silaging and milking cows for local farmers and in the winter laying hedges and selling logs. The build up of contract work was mainly by word of mouth, the distribution of personal cards and adverts in the local and farming press, the emphasis on friendly prompt advice and service. The money he earned bought the fencing materials for him to stock fence the farm, with both fencing wire and mains electric fencing. Little did he know at the time that fencing was to become a major business.

In the Autumn of 1983 Antony went to Dorchester sheep sale and bought 60 white face border Leicester, two tooth, ewes in order to produce good quality meat stock. With good Suffolk rams purchased locally and the original ewes with their ewe lambs from both 1982 and 1983 lambings, we were ready to lamb 180 ewes in the April of 1984. I was still a full time teacher; Antony put the rams in with the ewes in November, so that I could assist with the lambing during the Easter holidays.

During the winter of 1983/84, Ben, the sheep dog died of old age and 'Spot' was purchased from a local farm. She was six months old and very keen to be put to work. Antony and I both thoroughly enjoyed our first lambing together. Tending to the ewes and lambs was physical work but was well worth the reward, despite the lack of sleep. The expecting mums need to be fed with a food concentrate poured into troughs from 56lb bags, difficult to lift when you are tired and my size! They also require water and a constant check on them to identify when they are coming into labour.

If a ewe is having some difficulty delivering, assistance is required. First the ewe has to be caught, you would never know that they are about to give birth the way some of them can sprint out of your grasp! Once caught an internal inspection is done; Antony has taught me all I know and with my small hands I am more able to assist ewes having difficulty delivering.

I shall never forget the very first time Antony asked me to roll up my sleeves, on a freezing cold evening and insert my hand into the warm interior of a ewe, to find the front legs of the lamb who was trying to come out head first! I psyched myself up and was fine during the operation with the knowledgeable advice and encouragement from Antony. After the experience, I felt rather queasy but was quickly overcome with joy and exhilaration when I looked back at the little life I had possibly saved, which was being so lovingly cleaned by his proud mum!

It was in 1984 that the farm, with Antony's contract work, started to support itself, in terms of any other machinery and modernisation it required. In the early months of 1984 the farm purchased the Massey Ferguson tractor and sawbench to assist Antony's forestry work and sale of logs. In the summer of 1984 we bought a mower and the hay bailer that had been borrowed for the previous two years from a local farmer. In the winter a chainsaw was purchased to assist with the forestry work.

The quality of the grass in Hedge and Big fields was poor and even with a vast amount of fertiliser would never support a large number of sheep, so in the Autumn of 1984 they were ploughed and reseeded with grass, to produce a thin crop of rye grass the following year which has thickened ever since. The majority of the fields were all stock-fenced by now and the hedges have been left to grow up to produce timber for future years and shelter for the sheep from the cruel Atlantic winds and rain common during the winter.

The amount and variety of stock on the farm vastly increased during the year of 1984. Three Jersey heifers, Delta, Gamma and Gema were purchased to keep Beta company and to be reared until 18 months when they would be put into calf. Some chickens and ducks roamed the Cleave until the fox sadly found the ducks so the chickens were given a new home in the roundhouse and Alpha lived outside in the summer. With Antony's increase in contract work the ponies, Marcus and Rupert, were sold in 1983. The stable became Alpha's new milking parlour and home for the winter because the shippen had been gutted, ready to be converted into a self catering unit for visitors.

Two gilts occupied one of the sites and were fattened on the skimmed milk and the scraps, from the now growing farm holiday business. A pair of geese made their home in the other sty and produced a lovely flock of goslings in the very cold spring of 1984. My contribution to the livestock was a little toggenburg goat I had felt sorry for whilst at Taunton market with Antony selling lambs. She became a family pet, coming for walks around the farm with us until she died four years later. Another addition was a Charolais X Friesian Bullock, called Smokie because of her grey colouring, obtained while Antony was purchasing eleven calves from a local farmer, to be reared for two years for beef. This was not a very profitable enterprise because it coincided with the announcement of BSE, making the price of beef slump. Bimbo, a liver and white springer spaniel, was the third addition to the farm, a faithful pet and companion for Spot and everyone who gives him attention!

With the 180 ewes, their 300 lambs and the above mentioned animals, this year and the next six years must have been the time when Huxtable Farm had to support the most stock during our time of farming. In fact, during most of the summers, the fields seemed to turn from lush green grass into black, dung covered fields, owing to the vast number of sheep they had to support and with the lack of rain to wash it away.

On August 3rd 1985 Antony and I were married! We had organised all the catering for the reception and evening celebrations ourselves, Antony had made the wedding cake, as he had for his sister's and brother's weddings, and I iced it. In the evening there was a barn dance in the farm yard, whilst one of this year's lambs roasted on a spit over a large open fire. This was a big day in more than one way. Barbara and Freddie gave us the most wonderful surprise we could have imagined. Alan Gordon-Lee, our solicitor, stood up at our reception and proceeded to read a deed, in old English, that Antony and Jacqueline Payne had been given 40 acres of land, The Hedge and Big fields, from Barbara and Freddie Payne as a wedding present. After a wonderful honeymoon in the Lake District, we sold my little house in South Molton and moved into the Shippen, when the last visitors went home.

The variety of stock on the farm continued to grow during 1985. Rare breeds were introduced with the addition of a small flock of Hebridean sheep and ram, two Vietnamese Pot Bellied pigs and some Silkie hens. The Hebrideans used to get on with the lambing by themselves, partly because they were difficult to catch and identify when they went into labour. In fact, they used to just appear in the field with a tiny fragile looking lamb who in comparison to our main flock lambs were very much stronger in relation to how quickly they got up on their feet and gambled away.

The pigs were called Ming and Mong. Ming got pregnant so was given a new home amongst the hay in the big barn and Mong had the sty to himself. It was whilst I was feeding some tame lambs, who were also housed in the big barn, that I could hear some strange squeaking noises coming from Ming's direction. She looked asleep, so at first I thought she must be talking in her dreams but the noises seemed to come from her rear end! On investigating I discovered that there were two tiny little wet piglets scrambling around amongst the hay, then to my horror Ming rolled over in her sleep, right on top of them. Not being able to move Ming, I just hoped that the hay beneath her had saved the piglets. Now it happened that I was on my own at the farm, Antony was out contract silaging so not knowing anything about pigs farrowing, I tried to contact him with phone call after phone call whilst running up and down to check on Ming and piglets' progress. When I finally spoke to Antony, he informed me that pigs seem to go to sleep whilst farrowing and all I could do was to make sure Ming was surrounded with plenty of hay to cushion the piglets when she rolled over. I sat, watched and hoped this was not going to be yet another disastrous experience with a farrowing pig. Six piglets were born but only three survived the traumatic labour. Ming was old, surviving long enough to suckling her young until weaning. The young were sold and Mong was hired out as a stud until he too was sold.

These were the last pigs on the farm for a while because in 1986 the pig sty was converted into a family room for guests. At the same time Antony built the sheep handling and dip area followed by the large barn behind his present day tool shed. This large barn housed all the ewes comfortably in the winter and the handling system made life a lot easier when sorting out and drenching the sheep.

The Jersey calves bought two years ago came into season and were artificially inseminated with sperm from a Limousin bull. All fell pregnant and produced beautiful calves. Zeta had very small teats and was very difficult to milk so she and the calves were sold at market whilst Alpha, Gamma and Delta produced gallons of creamy milk throughout the summer of 1986. Unfortunately the new milk licensing laws in 1987 meant that we could not put our own jug of milk and clotted cream on the guests' breakfast and dinner tables, unless we obtained a milk licence. Financially not feasible, Zeta and Delta were sold in 1987. This left Alpha who continued to produce beautiful calves each year until her fatal calving in 1988 during which she fell ill with milk fever, calcium deficiency, having just produced another beautiful live calf.

In 1986 Antony's contract work had expanded from casual farm labourer to working for Devon County Council, steam cleaning lorries. A steam cleaner was purchased and a variety of cleaning jobs were undertaken. The money earned from this enterprise financed the construction of the silage pit, next to the cubicle shed. A hired digger dug as deep as it could go into the solid rock of what is now known as Silage Pit field. The first year it was used seemed to involve more time, energy and machinery than if we were making and storing bales of hay. The pit had to be lined with plastic silage sheets, the grass brought in on a dry day, packed tight with a tractor driving over it and then sealed tight to prevent any air getting to the grass, otherwise it rots and is useless as feed. Feeding by this method proved wasteful and time consuming so big bag silaging was tried the following year, with the bags being stored in the pit. A big bale silage wrap contractor was hired and all assisted with putting on and tying bags. Soon after the silage had been bought in and stored, a young friend of ours visiting thought it was great fun to throw stones onto the bales and watch Bimbo try to retrieve them. This meant that quite a few of the bags were punctured and needed taping up to prevent the air from damaging the silage. Bag silage has been made each year along with a few bales of hay. In the August of 1989 103 bales of silage were harvested from Big field between 1030 and 1800 hours.

I remember the 1986 lambing as being the year when I was advised not to assist ewes with their deliveries because I was three months' pregnant myself. The new ewes bought in the autumn had toxoplasma abortion and were producing still-born lambs. This disease can infect humans. Our vet. advised me to have blood tests which, thankfully, were negative and on October 14th 1986, our first child, Nigel Frederick, was born in North Devon Maternity Unit, Barnstaple. I returned to teaching at South Molton for only one term because I was pregnant again very quickly. I stopped teaching and with our second child on the way, Barbara and Freddie gave Antony and I the remaining 32 acres of farm land and agricultural buildings. On the 1st October 1987, Antony and I became self-employed business partners and took over the farm accounts.

Our second child, Natalie Mary, was born on December 28th 1987, like Nigel, in North Devon Maternity Unit. 1987 was quite a year! We had become land owners, were increasing the size of our family and Antony had started fencing on the North Devon Link road.

The Big Barn required attention in terms of being useful on the farm. The ceiling between the floors had developed some dangerously large holes and the roundhouse roof was about to collapse, so in March 1988 we applied to the Ministry of Agriculture for a 'Farm Diversification Grant' to convert the barn and roundhouse into guest house accommodation. On November 9th 1988 we received a letter informing us that our proposed plan had been accepted, starting from the date of the letter and ending on September 30th 1989. A mortgage was taken out and work started immediately.

In anticipation of obtaining the grant Antony had concreted the floor and boarded up the walls of the open shed and made it into his new toolshed during 1988. The basement rooms of the barn were stripped of their possessions and the floor was dug out in order to make room for the foundations. It was not long before the corrugated iron roof was removed and new timbers were in place ready to support slate tiles - a vast improvement on the previous roof!

At Christmas a booking was made for some visitors to be accommodated in one of the room in March 1989 - the pressure was on! Two days before the guests were due to arrive everything was complete except for the glass in the windows. These were fitted and the rooms were first occupied on March 24th 1989. Barbara and Freddie's new home, the top of the barn and roundhouse, was completed at a later date and they moved in on July 21st 1989.

Over the years during the winter months Antony had managed to make a large pile of timber from his logging days. It was during the very hot and dry summer of 1989, on the night of August 2nd, Barbara woke to the sound of West Buckland School's fire alarm. Looking outside at 3 a.m. she saw a large glow, lighting up the sky in the direction of where the logs were stored. We were alerted and on investigating we found 40 tons of Antony's hard labour ablaze! The fire brigade was alerted and arrived to control the fire. They used two and a half thousand gallons of water, obtained by an engine travelling to and fro to East Buckland crossroads and not one of the guests, staying at the time, heard a thing during these small hours.

Antony was now fencing away from home a lot; he took on a Youth Training Scheme (YTS) student, Phillip Beer, from Chittlehampton. Phillip started by assisting with the 1989 lambing then continued to work with Antony fencing on the Link Road. Phillip is a Devon lad who had never visited Exmoor or the coast until he met Antony. He used to pedal a bike to get to and from work, until he saved up enough money to purchase a small motor bike. Phillip stayed as a YTS student working with Antony for a year then left because he wanted more farm work which Antony was unable to offer him. Six months later Phillip approached Antony for work and has continued to assist him with his fencing and farm work to this day, as a self employed labourer.

Antony's contract fencing took off, working at least six, twelve hour, days a week. This meant that he required more powerful machinery. The first exchange was to give the ever faithful old Landrover an overhaul and obtain the Toyota truck, followed shortly by the purchase of the John Deere tractor in April 1989 having sold the Massey Ferguson. The Link Road was complete and open to the public in September 1990 and with a large fencing contract established at Challacombe Estate on Exmoor, a powerful post driver was purchased. This increase in outside work and my commitment to the holiday business, having taken over from Barbara in October 1990, meant that the farm stock had little attention. With the drop in sheep prices and ever increasing costs such as feed, drench and time consuming attention required to produce healthy stock, we sold 390 ewes and lambs in October 1990. We kept very old ewes to keep the grass down and because if they had gone to market would have only made about £5.00 each, just enough to cover the haulage to the market, lost us money if they had not been sold, and were still capable of lambing another year without requiring a lot of attention!

The 40 old ewes got on with lambing in the Spring of 1991 requiring very little attention. Antony and I had missed the lack of sleep and thrills of midwifery so much, that we purchased 67 Masham ewes from a local farmer, in November 1991, to increase the flock numbers. With the start of the recession in 1991, the large fencing contract on Exmoor came to an abrupt and bad financial end with Antony being owed a substantial amount of money.

With his machinery and guaranteed standard of work Antony has continued to get good fencing contracts to this day, including an increasing amount of fencing work for the Fortescue Estate.

To illustrate the changes that have taken place in the farming stock and their value during our few years of farming, here are our 1984 and 1989 records:-

1984

1989

Cattle:	1 Cow	@ £300
	3 Heifers	@ £300
	1 2yr old	@ £400
	1 Yearling	@ £150
	1 Calf	@ £100

Sheep:	160 Ewes	@ £50
	15 Ewe Hogs	@ £40
	6 Rams	@ £60

Pigs:	6 Slips	@ £25
	1 Boar	@ £25

Poultry:	6 Hens	@ £2
	2 Geese	@ £10

Total Value   £11017

Sheep:	180 Ewes	@ £30
	70 Lambs	@ £25
	2 Rams	@ £50

50 Tons Silage @ £20/Ton

Total Value   £8250

Our third child, Henry Benjamin, was born on February 18th 1990 and Evette Veronica, completed our family when she was born on May 22nd 1991.

The land at Huxtable has not only supported a variety of stock but also crops. In 1988 a new game keeper was employed by Castle Hill Shoot who rent ground from the Fortescue Estate. To develop and expand this shoot the game keeper asked if we were interested in providing cover for pheasants. So, in 1988 Silage Pit field was ploughed and directly drilled with swede and turnips, 1989 kale and in 1990 swede and turnips again. In 1991 Corner field was ploughed and directly drilled with kale and left to reseed its self. In 1992 whilst Silage Pit field was sown with maize (the children enjoyed hiding in) and Middle field housed Partridges in tent like pens during the summer of 1992. This means that within 13 years of selling Huxtable Farm, the Fortescue Estate, indirectly, rent back part of our land. The holiday business is now the main farm enterprise and with the ever increasing difficulties of agriculture, we are glad to have established the contract work and holiday business. For the future we wish to make better use of our farm land by putting half of it into hardwood woodland, to enhance the beauty of the farm and as an added attraction for our visitors.

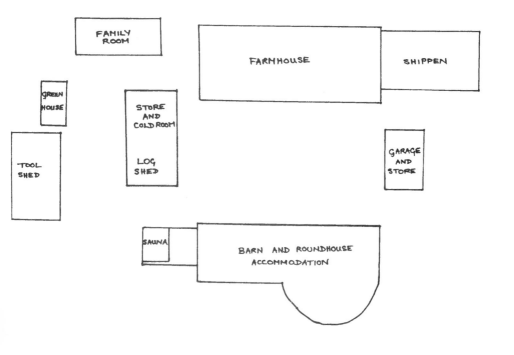

The present layout of the farm buildings

Chapter 9. SHEPHERDING AT HUXTABLE FARM

Our flock of sheep has varied over the years but have always been bred to produce a lean carcass demanded by today's consumers, and not for their wool. Antony started with a mixed breed of old ewes, who were cost effective when starting lambing. By selling the lambs the old ewes produced, we gradually acquired some white face Border Leicester ewes and some pure Suffolk rams. The offspring ewe lambs of these, were kept to build up flock numbers. Texel rams were introduced into the flock, so at one time we had a very mixed flock of Suffolks, Border Leicesters, Suffolk X Border Leicesters, Suffolk X Texels, Suffolk X Border Leicester X Texels. Some Hebrideans, a rare breed, who seemed to produce beautiful, strong, little lambs, joined the flock for a time. This entire flock was sold when Antony had a lot of outside fencing work on, and now we have a few Mashams, who make very good mums and seem to have very few problems during lambing.

Over the years Antony has used the original buildings and built other good, general purpose ones on the farm. The cubicle shed, with its three large bays, has one bay with individual pens used to house the ewe and her lamb or lambs, whilst they get to know each other. The other two bays are where expecting ewes are housed and fed during their pregnancy in the winter months. We keep our ewes indoors from about January until they lamb in March or April, letting them out depending on the weather and ground conditions. The idea is not as some may imagine, to keep them warm, but to keep them off the pasture. Our pasture is very well stocked in the summer and autumn, it does it the world of good to rest it during the winter. The combination of winter rains and the hooves of the sheep will turn pasture into what looks like a well used rugby pitch, the grass is worn away and the soil structure impaired. Wet conditions also encourage footrot, a bacterial disease of sheep which can make them lame, so bringing the sheep in has much to commend it. It also makes it easier to feed the sheep.

Many people do lamb their ewes outside, maintaining that it is healthier, the drawback as we have found out, is that if a ewe requires assistance during lambing, it is very hard to catch, despite the fact that she may have a lamb half born! We tend to bring the ewes in nearer lambing, having split them into groups according to their expected date of lambing. We feed ewe and lamb pellets (containing barley, wheat and protein fish meal), 1lb a head per day. A ewe is actually "in lamb" for 147 days but it is not until the last eight weeks that the foetus really begins to grow, and when a ewe is carrying twins or triplets, the stomach becomes too small to hold sufficient grass to sustain both the ewe and her lambs, hence the need to feed concentrated food.

In 1988 we ultrascanned the ewes about eight weeks before their expected date of lambing. The ewes were marked, according to how many lambs they were carrying and split into two groups, singles and ewes carrying two or more lambs. The singles were not fed concentrates whilst the ewes carrying two or more ewes were. By knowing the expected number of lambs each ewe was going to produce, not only cut feeding costs, but also meant that we were able to foster triplets on to ewes with single lambs. Not all ewes have enough milk to feed three hungry lambs, so if necessary one lamb is fostered onto a singlet ewe to even out the demand on the udders!

Fostering a lamb involves fooling the mother into believing the intruder is her own. We try to catch a singlet ewe whilst she is lambing, and make her believe that she has had twins instead of a single. We do this by pushing the hand inside the ewe for a time, to resemble giving birth to a second lamb, and smothering the fostered lamb with the ewe's water bags and membranes produced during lambing. Some ewes accept the fostered lamb straight away, others may need to be restrained for a few days whilst the new lamb suckles; by the time the mother is free to butt and kick it away, the lamb smells so thoroughly of her that she accepts it. Only in desperate circumstances, when we have had a ewe with gallons of milk, who has lost her lamb, have we skinned the dead lamb to make a jacket for the foster lamb.

During pregnancy the ewes need to avoid sudden, severe frights or pain, excessive heat or cold, loneliness or overcrowding, because these conditions can cause stress, which in turn can reduce the body's resistance to infection in certain stages of pregnancy, causing the ewes to abort. About two weeks before lambing, the ewes are injected with a clostridial vaccine, which will protect the lambs in their first 12 weeks of life via the mothers' colostrum milk, against lamb dysentery and tetanus. The rich colostrum, (the first milk from a ewe), also contains antibodies, vitamins, proteins and laxative properties, so its is vital that the lambs receive this milk in order to remain healthy.

As the lambing time draws near Antony steam cleans the cubicles and lambing shed and puts down fresh straw. We also try to collect together all equipment we may need. Surgical gloves and lubricating jelly. to help when one of us need to put our hand inside a ewe. Iodine, which we place on the born lamb's navel. This protects against jointil, a disease which affects the joints. The bacterium gets into the blood stream via the navel. We also keep a few bottles and teats, just in case we obtain a few orphan lambs, known as "tame lambs". This may be the result of a mother not being able to produce enough milk for her lamb, or if she has triplets, we usually take one of the lambs away having received is colostrum. Sometimes the lamb may be born too weak to suck from its mother, then we have to use a stomach tube to feed milk into its stomach. A supply of syringes and needles of all sizes are kept, along with a supply of magnesium solution, calcium and antibiotics. The magnesium is for magnesium deficiency, sometimes referred to as lambing sickness. Until now we have only had about six cases of calcium deficiency, usually because the ewes were carrying three or four lambs. The antibiotics such as penicillin are used to guard against infection and are administered when there has been a difficult lambing and we have to put our hand in the ewe to assist her, or a weak lamb requires some medication. A lambing lamp is also used to revive a weak lamb and keep it warm.

Once the first lamb appears. lambing has started and the ewes are constantly watched and attended to. Most of our ewes will lamb unattended. With a practised eye, we can spot which ewe is going to lamb next. Her udder will fill with colostrum, she will separate herself from the main flock, will be restless and start to scrape the ground with her foot, as if making a nest. She may 'baa' a lot, and grind her teeth. The ewe will keep lying down, then standing up again, then starting to strain. A waterbag or two appears and may rupture, followed, hopefully, by two forelegs and the head of the lamb. The ewe lies down and strains several times and with a slither and a bump, the lamb arrives, it lifts its head and shakes itself free of the membranes. The mother turns round to lick the lamb dry, baaing gently to it and the lamb responds by bleating back. It is always a beautiful sight to see the lambs lift their heads for the first time, to shake their long wet ears and take their first breath.

If during the lambing the two forelegs and head of the lamb do not appear first, this is known as a 'mal-presentation' and needs correcting. Examples of mal-presentation include two lambs trying to be born simultaneously, a lamb being born head first with one or both of its legs back, forelegs only present, and a breech presentation. A ewe lamb. lambing for the first time, can take up to twelve hours in labour, an older one can be in labour for four hours, so if lambing is taking longer than this, we may expect something to be wrong and will make an internal inspection. We may find a lamb in the breech position of that the ewe has what is called a "ringwomb", this is when the membranes have ruptured but the cervix has not opened enough for the lamb to come through. One or two fingers are gently inserted into the cervix and gently work it open.

If the ewe is carrying twins, after a short time she will abandon her first lamb, become restless and go through the straining process again. Once she has finished lambing, we check to see that both teats are producing colostrum. If there is milk in both teats the ewe is left to dry her new lambs and expel the placenta naturally. The lambs soon get up and try to find their way to the udder.

An hour or so later, the ewe and her lamb or lambs are then transferred to the individual "mothering" pens, in the cubicle shed, where the mother and new lambs get to know each other without interruptions. Sheep are inquisitive animals, and some ewes which have not lambed yet may push the new lambs around with their noses whilst others may "mis-mother", i.e. adopts every lamb they see as their's, licking and talking to it, making life very confusing for the true mother and its lamb.

Each pen has a bucket of water, bowl of concentrates and hay as a reward for the ewe. The lambs' navels are sprayed with iodine, rubber rings are put on their tails and the male lambs are castrated. The new arrivals and their mothers stay in their pens for at least 24 hours so that we can keep an eye on them to see if the mother is looking after her young and the lamb fend for themselves and take a drink every now and again. If all is well, the ewe and her lambs have the same number sprayed on their sides in order to identify whose lamb belongs to whom, and in good weather put outside to the fresh spring grass. There is nothing better than good, lush grass, to produce plenty of rich milk.

After turning out to pasture the ewes and lambs are constantly checked to ensure that all are healthy. Some common things to watch out for are mastitis in the ewes, diarrhoea, wet mouth and jointil in lambs. When sick, the ewes and her lambs are treated with the appropriate medication.

The shepherd's calendar at Huxtable Farm.

September: Buy new ewes and rams and keep separate until tupping time, (when the rams serve the ewes). Worm, unless told that they have been wormed recently. Vaccinate against clostridial disease, if bought in un-vaccinated. Trim feet if necessary. Dip, if necessary at least two weeks before tupping, the smell may put the rams off!

October: Feed the ewes and rams well, to get them into good condition, but not too fat, ready for tupping.

November: Put rams in with ewes. Inspect daily, record any marked and change raddle colour as necessary. (Raddle is the paint or crayon block harnessed onto the ram's chest, that leaves a coloured mark on the ewe's bottom once the ram has served, thus letting us know when the ewe was served and calculate her expected date of lambing. Give new ewes a second dose of vaccine if necessary. At the end of the month take out rams and keep in a separate field. (A ewe's menstrual cycle is three weeks, August to January, depending upon the breed, so any ewe served by the ram in the fourth week would be the ones missed in the first week of tupping).

December: May start to feed hay or silage, depending upon the weather and grass conditions.

January: Buy in straw for winter bedding in shed.

February: Worm ewes six weeks before expected date of lambing. Start feeding grass concentrate food six weeks before lambing; lambs grow fastest at this time and the ewe's stomach does not have a lot of space for hay or grass.

March: Vaccinate ewes with clostridial vaccine two or four weeks before lambing. Clean out lambing sheds. Collect together equipment required for lambing.

April: Lambing: 24 hours a day watch throughout (causes havoc with sleeping pattern). Worm and trim the ewes' hoofs. Castrate and docks tails of lambs.

May: Keep constant watch on progress of ewes and lambs. Shear ewes.

June: Shut up field or fields to be cut for hay or silage. Worm lambs at about eight to twelve weeks old (1st dose).

July: Dip to prevent maggots from blow fly. Wean lambs. Worm lambs at about twelve to sixteen weeks old (2nd dose). Worm ewes. Keep careful watch on udders of drying off ewes. Vaccinate ewes with clostridial vaccine, booster dose. Cut grass for hay or silage.

August: Sort out ewes into those who will be kept for breeding and those to be killed. Pick out lambs for market and slaughter house.

Chapter 10. HUXTABLE FARM HOLIDAYS

In 1980 Barbara and Freddie had the opportunity to sell their family home in Horwood, near Bideford, and to purchase Huxtable Farm, a small loss making sheep farm, to be run by their son Antony who had just finished an agricultural course at Seale Hayne College. I will endeavour to describe the transformation of the farmhouse, previously owned by an elderly and sick farmer, Fred Ridd, who sadly died of farmers' lung within a year of selling.

Barbara and Freddie invested all their capital into the farm, but because of various financial set-backs, had to take out both a mortgage and a bank loan. However, both Barbara and Freddie were still working and earning salaries and it was these salaries that provided the capital for all the initial considerable modernisation and restoration that was to be done to the property.

When I met Antony and first visited the farmhouse, a large amount of restoration had already taken place, as Barbara has described in her account of the first two years at Huxtable. The ground floor had the lovely dining room with the oak beams and screen panelling exposed, as you see it today, with a glowing warm fire burning in the fireplace. There was a large kitchen through the two doorways off the dining room and the present living room, a separate television room (once the downstairs bedroom for Grandma Ridd) and a utility room. Upstairs were four bedrooms and two bathrooms. One bedroom was occupied by Barbara and Freddie, with a bathroom backing on to it (once a small bedroom). Jenny, Antony's sister, had the "pink" bedroom and shared Barbara and Freddy's bathroom. Antony occupied the "blue" room, which was the bathroom during Doreen and Fred Ridd's time at the farm, and Barbara and Freddie renovated it into a bedroom with a shower and basin in an alcove (part of the disused chimney). These left the "yellow" bedroom with its own bathroom as a visitor's bedroom. Again, a lot of work had taken place upstairs, because in Doreen and Fred's time there was a large landing from the main staircase to the back stairs, which was now the bathroom to the yellow room and a separate toilet.

As Barbara has mentioned, in order to help with the finances of the farm, a Bed and Breakfast (and Dinner) business was started, with all the family contributing, because Barbara and Freddie were still working full time in very demanding jobs. In the spring of 1983 advice was taken from ADAS and our accountants which finalised the decision to develop the tourism side of the farm, aiming for an "up-market" trade with a longer period of full time occupancy. The accommodation was inspected by the English Tourist Board and Huxtable Farm became a member of the Farm Holiday Bureau, through an active local farm holiday group with similar interests. The farmhouse holiday trade developed fast and profitably through advertising in professional magazines, the English Tourist Board, local tourism offices and the Farm Holiday Bureau. Huxtable Farm also became a member of the Caravan Club for a few years. Campers and those in caravans enjoyed deciding where to set up camp, having 80 acres of land to choose from.

When Barbara retired in 1984, with more time to devote to the holiday business, the shippen was converted into a two bedroom self catering unit/bed and breakfast annexe. It was part of the holiday business until Antony and I were married in 1985 and moved into the shippen at the end of the holiday season. Most of the trade by now was by "word of mouth" or by visitors returning, the emphasis being on friendly and "open house" hospitality, with the guests having the freedom of our home and farm. The dinners were, and still are, candle-lit, of *Cordon Bleu* standard, using farm and local produce served with a glass or two of Freddie's home-made wine. Some Huxtable Farm recipes are given in Appendix 8.

Other than sheep, wide variety of animals were now kept on the farm solely as pets and for tourism interests! Three Jersey cows produced rich creamy milk, served fresh on the breakfast table, whilst the remainder was made into clotted cream so that there was always a large bowl of clotted cream on the sweet table in the evening. Pigs were kept and fattened on the skimmed milk, the by-product from the clotted cream production. Various breeds of sheep, ducks, chickens, geese, goats, ponies and dogs have all lived a happy life, and some still do, as pets at Huxtable. Unfortunately, due to the Milk Licensing Law of 1987, we had to cease putting our jug of milk and clotted cream on the dining table unless we obtained a licence, which was financially not feasible, so the Jersey cows were sold and subsequently the pigs, a sad end to an era!

In 1986 Barbara won the English Tourist Board's "Taste of Exmoor" competition, for her Whortleberry Cream dessert. The judge was the BBC Breakfast TV chef, Glynn Christian, who said that it was the only entry that contained the *gosh* factor! Barbara's prize was a £100 cheque and a beautifully engraved pottery bowl.

1986 was also the year then Antony and I started our family, with the arrival of our first child, Nigel Frederick on October 14th. That winter the pig sty was converted into a family room and Freddie moved his small fruit and vegetable plot to establish his present extensive fruit and vegetable garden in the small paddock known as "Pump Splat field", in order to supply the farm holiday business. During the summer months that have followed, jams have been made from the mountains of fruit the garden produces and many hours have been spent preparing vegetables for the freezers, to meet dinner needs during the winter months. Any surplus fruit has been made into Freddie's wine.

By now the holiday trade, when fully booked, was catering for seven adults and two children, on a dinner, bed and breakfast basis. This was too much work to be managed by the family alone, so we employed some local domestic help and student labour in the high season. We had Oxbridge graduates and girls from Japan and Germany, a male engineer and many others. The contrast between University life and North Devon must have proved too great as many left before the end of the season.

In 1987 our second child, Natalie Mary, was born and I stopped teaching, to be with the young family and assist with the farm holiday business. I purchased a word processor, with which I taught myself to become familiar with such terms as "files in limbo", and "whether a disc is formatted or not". Having grasped the basics, I prepared the farmhouse brochure information ready for the printers and produced the "Welcome" letter found in guests' rooms.

Barbara, now Chairman of the local farm holiday group, along with other Farm Holiday Bureau members in Devon, initiated the production of the *Devon Farms* brochure. She organized its press release at Bicton College of Agriculture in 1988. This event emphasized that the Farm Holiday business had become a professional enterprise compared to the initial idea of farmers' wives providing bed and breakfast in order to earn a little money for themselves.

With more and more families visiting the farm during the summer and our own growing children, a play area with swings and sandpit was developed in the paddock opposite the shippen. The Big Barn opposite the farmhouse required attention in terms of being useful on the farm, so in 1988 Antony and I applied to the Ministry of Agriculture for a "Farm Diversification Grant" to convert it into guest accommodation. We received notice saying that we were eligible for the grant and so Antony and I began our own farm holiday accounting. The first guests occupied the Barn on 24th March 1989 and in 1990 a sauna and shower room was built adjoining the rooms. By making these rooms *en suite* and with all the other restoration that had taken place over the years, the accommodation had now been brought up to English Tourist Board "Three Crown" standard.

Our third child, Henry Benjamin, was born on 18th February 1990 and with the holiday business now the major enterprise on the farm, Barbara, Freddie, Antony and I became partners. Barbara and Freddie have since become semi-retired, although they still play a major role with the production of fruit and vegetables, jam and wine making, and upkeep of the gardens and are always ready to listen and give their knowledgeable advice and opinions when required. The holiday business became V.A.T. registered, now that Antony and I were partners and V.A.T. registered with the farm. The bed and breakfast prices, unfortunately, were increased to make allowance for the V.A.T.

Opening all the year to visitors meant that we required adequate heating during the winter months, so oil fired central heating was installed, with the wood burning stoves boosting the system when lit.

The terrific winds in the February of 1991 that caused so much damage throughout the south of England took its toll at Huxtable. Slates flew off the ends of both the farmhouse and the converted barn, the electricity went off (quite a common occurrence during the winter), the stove in the farmhouse kitchen blew out, sending soot everywhere and then to top that mess, with an enormous bang and sound of smashing glass, we discovered on entering the kitchen that the roof had blown off the "green" room annexe, smashing into the kitchen window and spraying glass everywhere and opening the kitchen to the elements of the raging wind. Nothing could be done during the storm, it was far too dangerous to enter the kitchen and venture outside. When the storm had eventually passed, the damage was assessed. The tiles were replaced, new glass was put back into the kitchen window and the "green" room tarpaulined until the, now very much in demand, builders could put up a new roof and replaster the ceiling.

We recovered from this start to the year and the holiday trade continued to flourish. We had been recommended and entered into the **Which Good Bed and Breakfast Guide** and Elizabeth Gundrey's *Staying off the Beaten Track*, both very commendable publications. The English Tourist Board introduced a quality grading system and Huxtable was graded, and still is, "Commended".

With our fourth child due in May, our only baby to be born during the peak season for visitors, Antony and I trained some students to take over whilst I was out of action. Two weeks before the expected date of delivery an incident occurred and the students were asked to leave. These two girls were the last of the many "summer" students who had been employed. Panic struck! Thankfully a local friend came to the rescue and I learnt to cope with P.A.Y.E.

Evette Veronica, our fourth child, was born on 22nd May 1991, making our family complete.

Having a great interest in the conservation of this beautiful countryside we live in and, being keen to encourage visitors to enjoy and appreciate the area, I became a member of the *Tarka Country Tourist Association* working party. I assisted with the establishment of the Association which was set up in February 1992 and continue to enjoy being an active member of the *Tarka Country Tourist Board Association's* committee, promoting the countryside and the Tarka Trail which passes the entrance to the farm.

1992 has been quite a year to remember. In the spring, Huxtable hosted a trial Tourist Board inspection of our local farm holiday group members, a valuable experience and a wealth of information was gained by all. During the lambing season in March, a Belgian film crew descended on us to make a film for a holiday programme broadcast in Belgium. The film was based around Agatha Christie with a lot of the presentation shots in and around the farmhouse. This was a very interesting experience for ourselves and our visitors. A video of the programme was given to us thanking us for our hospitality. Huxtable was registered with Devon County Council as food operating premises when the new Food Hygiene Law was introduced and we have attended food hygiene courses to obtain certification in food hygiene. The amount of waste that accumulates with the holiday trade is astounding, so we try to collect the newspapers, tins and bottles and take them to South Molton's recycling centre. With this interest in recycling we entered Devon County Council's *Recycling County Tourism Award*, and were one of the six in the county commended for our efforts. I attended an advertising course in the autumn and consequently rewrote and designed our new colour brochure. A facsimile/telephone answering machine was also purchased, in order to continue the professional growth of our business.

During our short time at Huxtable we have experienced many happy, enjoyable moments and during our trading years made some extremely good friends and met very interesting and charming people from all over the world. With any business, though, we have also encountered some very strange situations.... but that is another story.... or even another book!

Appendix 1. THE DEVON MUSTER ROLL for 1569 (i.e. those who bore arms, not protestors).

Hundred of Braunton.

Coombe Martin Parishe:	Archers:	11
	Harquebusiers	16 (including one ROGER HUCKSTAPLE)
	Pikemen	6 (including ANDREW LOVERINGE)
	Billmen	3
Berry Narber Parishe:		None
East Bucklonde Parishe:	Archers	4 (including one RICHARD HUXSTAPLE)

Hundred of Shirwell

Charles Parishe:

Presented JOHN HUXSTAPLE (in possessing goods valued £10 and £20)

    1 Bow
    1 Sheaf of arrows
    1 Steel cap
    1 Bill
    Billmen   3 (including JOHN HUCKSTAPLE & CHRISTOPHER HUCKSTAPLE)

Appendix 2. DEVON PROTESTATION RETURNS for 1641 (i.e. protesting against bearing arms)

Hundred of Shebbear

Buckland, Filleigh Parish	Richard HUCSTABLE
	Wm. do.
	Abraham HUXSTABLE
	Clement HUXTABLE
	James do.

Hundred of Braunton

Barnstaple Parish:	None
Berrynarbor do.	None
Bratton Fleming	Wm. HUXTABLE
Braunton	Henry HUXTABLE
do.	do. do.
East Buckland	John do.
West Buckland	Thomas HUCKSTABLE
Coombe Martin	Richard HUCSTABLE
Heanton Punchardon	John HUCSTABLE

Hundred of Fremington

Newton Tracey Parish:	Nicholas HUXTABLE
Tawstock	Richard HUCKSTABLE

Hundred of Shirwell

Arlington Parish:	John HUCKSTABLE
Charles do.	1. Christopher HUXTABLE
	2. Christopher HUXTABLE
	3. Christopher HUXTABLE
	Edmund do.
	George do.
	John do. (signed *overseer*)
	John (Senior) do.
	John (Junior) do.
	Richard do.
	Richard (Junior) do.
Highbray Parish:	Wm. HUXTABLE
Lynton do.	David HUCSTABLE
	Amos do.
	Antony do.
	George do.
	John do.

Hundred of North Tawton
Dowland Parish: William HUXTABLE
Bishops Tawton do.   Davy HUXTABELL
South Molton do.   Michael HUXTABLE

Appendix 3. WEST BUCKLAND CHURCH

Inscription of a slate gravestone immediately outside the main (south) entrance to the Church.

On east side of stone - *This stone is inscribed to perpetuate the memory of the last of the HUXTABLE'S of Leary in the Parish of Chittlehampton, who have been Land owner,s of parts of LEARY before the days of OLIVER CROMWELL, the latter part of whose LIFE was devoted wholly to the service of the Saviour. His end was peace.*

(i.e. before 1600) (N.B. The commas are as shown)

On west side of stone - *In memory of BETTY Daughter of JOHN and Betty HUXTABLE of the Parish of Chittlehampton who departed this life on the 7th Day of March 1806. Aged 4 years*

Also ELIZABETH daughter of the above JOHN and BETTY HUXTABLE who departed this life the 18th day of December 1808. Aged 9 months

Also RICHARD son of JOHN and BETTY HUXTABLE above said who departed this life on the 25th day of April 1811. aged 2 months.

Also in memory of BETTY wife of the above JOHN HUXTABLE who departed this life on the 25th day of September 1836 aged 58 years.

Also in memory of JOHN HUXTABLE husband of the above BETTY HUXTABLE who departed this life the 13th day of October 1857 in the 84th year of his age.

Appendix 4. WILL OF THOMAS HUXTABLE

Yeoman of East Buckland: 11 July 1622.
Proved 23 Aug. 1622 by Katherine Huxtable, oath coram
Bartholomeo Moore, rectore de Highbray.

To my son Thomas H. my lands & tenements in Heanton Punchardon,
Braunton, South Molton & Marwood; also £50.

To my daughter Rose H. £65:13:4: within 3 years, or within 2
years of marriage.
To my daughter Mary Widlake.
To my daughter-in-law Joan H.
To John H., son of my brother John H. the rent which my sd. brother
oweth me for Newport close.
To my sister Anne Greene.
To Thomas & Richard sons of Richard H. decd. at 21.
Residue to Katherine my wife whom I make my exx.

Overseers, my brother-in-law, Mr. Bartholomew Moore, Mr. William
Pyncombe & Richard Holway.
(W) Bartholmewe Moore, clerk, Hawle,
   Willm. Pyncombe, Richard Holwaye.

WILLS: HUXTABLE family. (N.B.) Devon and Somerset Wills were
destroyed by bombs in 1942.) The following are listed without addresses.
If the wills had been available I am sure that they would have contained
useful information about property.

1602 27th November: JOHN of CHARLES. (Surname spelt
HUCKSTABLE). Wife: JOAN, who proved the Will. Daughters: Ellinor,
Elizabeth and Anne. Sons: Richard and Edmund. (Some left to poor of
Parish).

1622 11th July: THOMAS of EAST BUCKLAND, YEOMAN (Surname
spelt HUXTABLE). See above.

1624 30th December: RICHARD HUXTABLE (as spelt) of Walland in
Charles (Husbandman). Proved by ANNE, Wife. Sons: Richard (eldest),
Edmund. Mother: Jane. Daughter: Elizabeth. Some given to poor of
Charles.

Analysis of WILLS - Barnstaple Register 1563 to 1858.

Name: HUXTABLE

Recorded at:	1500-1600	1600-1700	1700-1800	1800-1900	Total
Ilfracombe		2	3	7	12
Berrynarbor				1	1
North Molton	1	1	8		10
Chittlehampton		2	7	7	16
E. Buckland	1	1		2	4
Buckland-Filleigh		2	2		4
Charles	2	14	6		22
Bratton Fleming	2	2	4	1	9
Combe Martin	1		1	1	3
South Molton		1	3	8	12
Lynton		1		1	2
Barnstaple				1	1
Warkleigh		4	2	2	8
Burrington			1		1
West Down		4	1		5
High Bray		2			2
Challacombe		3	1	1	5
Bulkworthy	1				1
Arlington	1	1			2
Bondleigh				1	1
Marwood			1		1
Shirwell			1	1	2
Tawstock		2			2
High Bickington			1		1
Northam		2			2
Coleridge			1		1
Fremington			1	1	2
North Tawton			1		1
Clovelly			1		1
Goodleigh		1			1
Roborough	1			1	2
Instow				1	1
Dowsland		1			1
TOTALS	10	46	46	37	139

Incidence of spelling:

HUXTABLE	5	33	36	37	111
HUXSTABLE	3	13	9	-	25
HUXSTAPLE	2	-	-	-	2
HUSTABLE	-	-	1	-	1
TOTALS	10	46	46	37	139

NOTES

1. It was observed that the nearest to the original spelling, viz: Hokestaple occurred twice only and then in the earliest group (1500-1600) so that the affinity was declining until 1800 after which the present spelling was complete and total.

2. It was also observed that those two nearest-original spellings occurred at EAST BUCKLAND and at CHARLES. This goes a long way towards the proof that the name originated in East Buckland and changed as the "family" drifted towards the Northern areas of Charles and Ilfracombe (including Berrynarbor).

3. C. Spiegelhalter notes: The "de" has disappeared. About 1318 to 1329 the particles "de", "le" and "atte" began to be dropped. Until the end of the thirteenth century few surnames other than landed classes had been hereditary.
Personal note: This is an interesting step in the transition to the present spelling.

4. Dr. W.G. HOSKINS notes: "Where a distinctive SURNAME is found over centuries in one place in successive taxation returns and the like, family CONTINUITY CAN BE PRESUMED even if the descent cannot be proved step by step".

Appendix 5. ARCHIVAL NOTES

Notes from research at the Devon Studies Section of the EXETER Library in order to assist in tracing the origin of the farm and the family.

1330 Subsidy Roll
John de Hokestaple, East Buckland (N.B. This roll is not in the Exeter Library but I believe may be at Barnstaple. Between 1290 and 1334 seventeen Lay Subsidies were levied.)

1332 Devonshire Lay Subsidy Roll
EST BOKLAND - 10 people listed
WEST BOKLAND - 10 people listed
BOKLAND DYNHAM - (i.e. NORTH BUCKLAND) - 8 people listed

1337 MAP of HUNDREDS shows that between the River TAW and the DEVON Boundary on EXMOOR there were three HUNDREDS viz - BRAUNTON (roughly from Ilfracombe to South Molton).
SHIRWELL (includes the Berrynarbour and Lynton area).
MOLLAND with NORTH and SOUTH MOLTON.

1216 to 1649 INQUISITIONS, POST MORTEM (Devon and Cornwall Record Society). No Huxtables or derivatives are shown.

1196 to 1272 DEVON FEET OF FINES (Richard 1 to Henry 111). These are records of change of ownership or the end of a court case to determine ownership. Hence finis or finishing-off. Might have started with ROMAN law in the twelfth century, but probably after the Conquest.
1244 BERY NERVERD (Berrynarbour). No Huxtables or derivatives.

1272 to 1369 FEET of FINES (Edward 1 to Edward 111). 1357 Christian names - JOHN, WILLIAM and ISABELLA appear. (Surnames to be checked). 1320 Berrynarbour spelt BIRIENERBERT. Also in Devon & Cornwall Notes and Queries Vol.1, page 240, spelt BURYN ARBER.

1402 JOHN HOKESTAPLE. Card index states "pardoned of outlawry for not appearing to answer ALAN HEDDON touching a trespass". Vide Calendar of PATENT ROLLS page 79 in the main library. This actually states: "May 10th 1402 (HENRY 1V - Pt. 11) Westminster. JOHN HOKESTAPLE, for not appearing to answer ALAN HEDDON touching a trespass... DEVON". There is no other detail.

1524 to 1527 Devon Subsidy Rolls:
EST BUCKLAND: THOS HUCSTAPULL and sixteen others who paid tax;
WEST DOWNE: JOHN HUCSTAPYLL; CHARLES (spelt: CHARELYS):
ROBERT HUCSTAPULL. (N.B.) These are the only ones of the family name
in Devon for 1524 to 1527).

1540 (circa) HUKESTABULL JOHN: monastery prisimir (sic) employed at
EAST BUCKLAND by ESINORA PERCKHAM, widow. (Devon and Cornwall
Notes and Queries XV11 page 240.) This record occurs in lists relating to
persons ejected for various reasons e.g. "because he is a fermer" etc.)

1196 to 1272 Devon Feet of FINES
1272 to 1369 Devon Feet of FINES
No mention of Huxtable or derivatives in BOCKLAND, EAST BUCKLAND
or WEST BUCKLAND.

DEVON RECORD SECTION - EXETER LIBRARY - 12th December 1988
Notes made from microfiche.

Part I WEST BUCKLAND PARISH REGISTER 1625 to 1677
All hand written and in poor condition. Difficult to read and a lot missing.

18th My 1680 Katherine Braylye m James Buckingham.
1664 James Finning (?) of Bartholomew (house?) & Elizabeth his wife.
Part II (Most of the foregoing is illegible)

BAPTISMS 23 Dec. 1686 Son of Henry Huxtable (sic). Christening of
Chittlehampton.
1695 23 Apr Henry son of Henry Hocstaboll (sic) of Chittlehampton.
1698 17 July Huxstboll (sic) son of Henry & Elizabeth his Wife of
Chittlehampton. Baptized.
1731 7 Mar Henry son of James Buckingham and Susanna, Wife. Baptism
date missing. Huxstable (sic) of Chittlehampton.
1744 Joan Daughter of Hy Huxstable of Chittlehampten. Baptism.
1760 Elizabeth Dr of Rd. Huxtable (sic) & Mary wife Chittlehampton. 24 Nov.
1770 Wm. Buckingham signed as Curate.
1771   ditto.
1783 June Baptism Thos son of Wm & Grace Buckingham
1794 Grace Dr of Edmund and Joan Buckingham. (October)
1797 Buckingham (illegible). Several (three at least) Buckinghams. Baptisms.
1792(?) 1892 Buckinghams. Several Baptisms.
1801 Huxtable (sic) 8 Feb 1802(?) Baptism
(illegible almost). Several Buckinghams again.

1805 Buckinghams.
1806 Maria Dr of John Huxtable (sic) & Betty Wife. 24 Jy 1806.
1808 Apr 21 Elizabeth Dr of John and Betty.
1810 Buckingham
1811 Rd Son of John & Elizabeth Huxtable (sic) of Chittlehampton.

MARRIAGES WEST BUCKLAND PARISH
1772 16 Oct. John Selly & Grace Huxtable (sic) Husbandman. West Buckland.
1760 3 Feb John Fairchild & Elizabeth Huxtable (sic) Yeoman. W. Buckland.
1780 (circa) Several Buckingham marriages.

BAPTISMS 1817 onwards. Buckinghams prolific.
1825 26 June James son of James and Mary Huxtable (sic) Chittlehampton.
1826 17 Jan Sophie Dr of Rd & Joan Huxtable. Chittlehampton. Farmer.
1826(?) 1825 17 Jan Wm Huxtable son of Rd. & Joan Huxtable (sic) Chittlehampton. Farmer.
1826 17 Jan Mary Jane Dr of Rd and Joan Huxtable (sic) Chittlehampton. Farmer.
1826 17 Jan Lydia Sarah Dr of Rd & Joan Huxtable (sic) Chittlehampton. Farmer.

BURIALS
1792 June 2 Margaret Huxtable (sic).
1808 4(?) Mar Burial. Elizabeth Huxtable (sic).
1806 21 Oct Burial. Rd Huxtable (sic).
1811 1 May Rd Huxtable (sic). Chittehampton.

Appendix 6. SMALL FARMING.

From the *North Devon Journal of 1883*

There have been of late a great many questions put to different authorities upon dairy farming. Persons asking what so many acres would return or cost to cultivate. The answers have been few; but one person asks the probable income from a 75 acre-farm. The writer considers that arable will yield quite as much as grass, and assumes that all it grows is consumed, or at least its equivalent. He puts the rent at £150; the stock - a bull, twenty eight cows, and two horses. Milk-selling being the business he omits pigs. Payment to out-going tenant, £100; stock, £665; cows costing £20 each and horses £35 each; implements, £145, or £12 an acre; working expenses, £440 per annum, including rates £20; labour, £120; cake, &c., £100. This labour is for two men at 21s., and 15s., and a boy at 6s. per week. Each cow is assumed to produce 600 gallons which by-the bye, is 7½ quarts a day for 45 weeks and this is to sell at 8d. a gallon. Four heifers are to be annually saved for the dairy, the remainder selling 50s. at a week old. Thus the calves are to produce £60, the milk £552, having deducted that used for the calves. The total receipts would be £612, showing a gross profit of £152, or, less interest at 5 per cent., £127, which is not a very large sum. This shows no provision for losses of any kind, and if cheese were made the profit would be less. Regarding the above statistics, a writer says, "It may be observed that £2 is too high a rent per acre - if I may be guided by the scores of admirable farms known to me - and that £20 each one is not nearly enough for cows giving 600 gallons; how many farmers are there who would travel far to find such cows at the price? £100 should be enough for implements until the farmer feels his way, for mere milk-selling demands few articles. Again, good land should carry more than twenty-eight cows to the seventy-five acres, especially as £100 is spent in cake. A really hard-working, astute dairy farmer, thoroughly understanding his work, would do this easily, and rear more than four head into the bargain, and at the least he should reap £150 a year, or £2 an acre, after deducting interest and allowing a sum for losses."

8: July 1807.— A Particular and Valuation of Huxtable Estate in East Buckland Devon.

Names of the Fields or Parcels of Ground.	Computed Number of Acres	Value p Acre	Amount p Annum.
	a	s.d	£.s.d
Dwelling House Out Houses Cowlodges and Green or Plantation	2		
Higher Garden	0¼	42.	1-1-0
~~Lower Garden & Orchard adjoining~~	0¼		
Mocks Close	1	28	6-6-0
Furzey Cleave in Do.	2	7	
Slade ~~Furze in~~	4	26	5-3-0
Furze in Do.	1.	7	
Cleve under the Slade	0½	14	0-7-0
Hill Close (in Wheat ye First Crop)	5	26	6-10-0
Mowhay adjoining	0½	26	0-13-0
Great Bottom Close (Barley 2. Crop)	5½	26	7-3-0
The North Little Bottom Close	3¾	26	4-17-6
South Little Bottom Close (in Barley ye 2 Crop)	3¼	23	3-14-9
South Down	4	22	4-8-0
Lower South Down	4	16	3-4-0
South Hill otherwise Furze Close	4	15	3-0-0
Coppice in Do.	0½	7½	0-3-9
Great Orchard	2	42	4-4-0
The Ham	2	28	2-16-0
Coppice above the Ham	4¼	7½	1-12-0
Coppice at the Bottom of Great Orch.	1	7½	0-7-6
Mocks Meadow	3	42	6-6-0
The Down (in Oats Third Crop)	11	17	9-7-0
Easter Higher Park	4¾	28	6-13-0
Wester Higher Park	3¾	28	5-5-0
Meadow adjoining the Green	0½	42	1-1-0
Pond Meadow	1½	50	3-15-0
The Hease	0¾	50	1-17-6
The Great Meadow als North Close	3¼	40	6-10-0
Copice under Do.	1	7½	0-7-6
	83	----	96-12-6

	£.s.d		
Deduct Land Tax	4-4-0		
Do. 120 Poor Rates at 8d pr Rate	4-0-0		12-4-0
Do. Repairs to be done by Tenant	4-0-0		
Clear Yearly Value			84-8-6

Appendix 7. HUXTABLE FARM MAPS AND WALKS

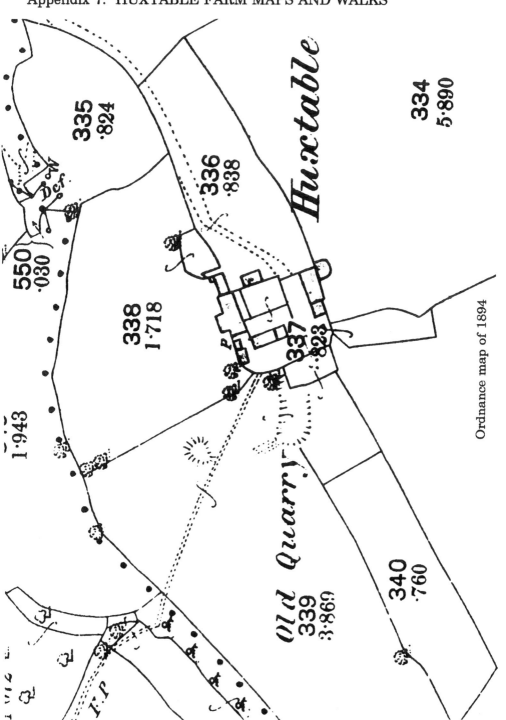

Ordnance map of 1894

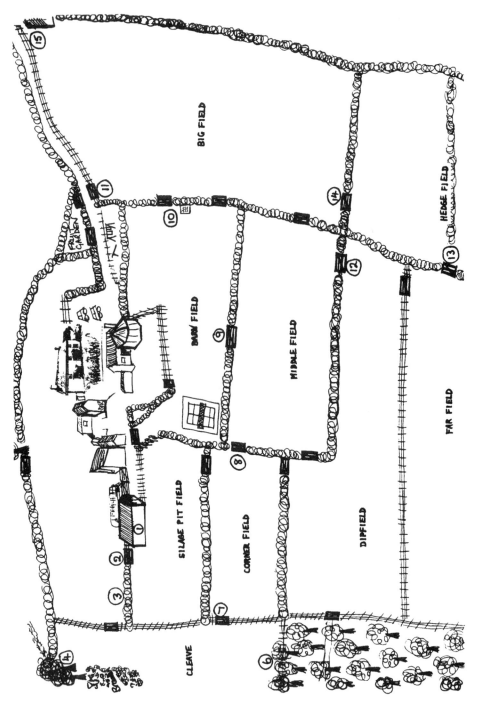

Map of Huxtable walks

Whilst walking please remember to shut any gates that were shut when you came to them and enjoy yourself!

Short walk: approximately 1 mile. Long walk: approximately 2 miles.

1. Start at the barn you find at the end of the lane.
2. Walk through the barn and go through the gateway on the right. In front of you, you can see West Buckland Village and to the right our sheep handling / dip area.
3. Follow the hedge on your left and go through the gateway into the Cleave.
4. Go down the slope to the right and through the trees to the stream. (Growing here are bluebells in spring and a large variety of fungi in autumn).
5. Follow the direction of the stream along the valley to the woodland.
6. Walk up the slope next to the woodland and along the top of the cleave to the hunting gate into Corner field.
7. Walk straight across Corner field to the gateway into Middle field.
8. *For the short walk*: follow the hedge on the left to the gateway, 9, into Barn field where the tennis court is. On top of the hill in Barn field near the small reservoir, 10, is the highest point on the farm (700 feet, 213m above sea level) with panoramic views of Dartmoor to the south west and Exmoor to the north east. Go through the gateway into Big field and walk down the field to the gateway, 11, into the lane leading back to Huxtable.

For the long walk: cross Middle field to Dip field, 12, walk up the slope to Hedge field 13, where there as an old hedge children love to climb.
To return to Huxtable, walk down the slope into Big field, 14, and up to the top of the field from which panoramic views of Dartmoor and Exmoor can be seen. Walk to the black shed and gateway, 15, into the lane leading back to Huxtable.

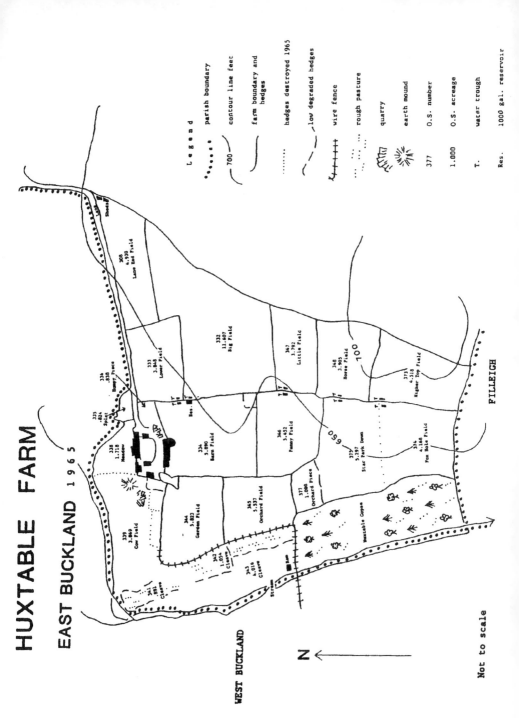

The Ridd's farm map

92

Appendix 8. HUXTABLE RECIPES

Many guests who stay at Huxtable Farm ask for our recipes. Here is a selection of those that are the most popular, starting with a soup recipe, for soups are favourite "starters" at the farm.

Huxtable soup

1 chopped onion
1 large peeled potato
1½ pints stock

Sweat the onion by lightly frying on oil but do not allow it to turn brown. Add the potato, stock and flavouring ingredients. Brings to the boil and simmer for 20 to 25 minutes. Process or liquidise.
Serve with croutons and warm bread. Serves 4.

This is the base for the majority of our soups to which a flavour is added, such as lettuce and sorrel leaves to produce lettuce and sorrel soup. Carrot and a little Madras curry powder to make a mildly curried carrot soup.

Mixed Herb Soup

This is a refreshing fragrant thick soup using home grown herbs.

50g (2 oz) Mixed Herbs. i.e. 8 large Sorrel leaves
a handful of mint and lemon balm leaves.
1 medium potato, peeled and diced
1 medium carrot, peeled and diced
1 medium onion, peeled and diced
900ml (1½ pints) water
450 ml (¾ pint) fresh milk
25g (1oz) plain flour
2.5ml (½ tsp) soy sauce
salt and pepper
small fresh mint leaves to garnish

Put the herbs, chopped vegetables and soy sauce into a medium saucepan with 900ml of water and boil for ½ an hour to reduce the liquid to 568ml (1 pint)
Puree the soup with the plain flour in a blender or food processor.
Reheat gently adding the milk, allow to thicken and season to taste.
Serve hot with three small mint leaves, homemade croutons and granary breadrolls.

Wild Mushroom Soup

This is a thick creamy soup with a strong flavour of the country.
In this menu I used the first crop of mushrooms but it can also be used again in the Autumn.

250g (8 oz) Wild Mushrooms, sliced or bought mushrooms
1 small onion, chopped
25g (1oz) West Country Butter or margarine
450ml (¾ pint) water
25g (1oz) plain flour
568ml (1 pint) fresh milk
Salt and Pepper

Melt butter in a pan and lightly fry the onions and then the mushrooms for 2 minutes.
Remove a tablespoon of lightly fried mushrooms and save for use when garnishing soup.
Place the mushroom mixture into a saucepan with the water and boil for ½ an hour until the liquid is reduced to approximately 300ml.
Puree the soup with the flour in a blender or food processor.
Reheat gently adding the milk, allow to thicken and season to taste.
Serve hot garnished with the lightly fried mushrooms, accompanied with home-made croutons and granary bread rolls.

Venison Casserole

700g (1½ lb) lean Venison, cut into 2cm cubes.
30ml (2 tbsp) sunflower oil
50g (2 oz) well seasoned plain flour
1 onion, chopped
2 carrots, chopped
1 large potato, diced
100g (4 oz) swede, diced
2 sticks celery, chopped
8 fl oz fresh milk
468ml (1 pint) water
3 x 10" spikes Rosemary

Heat the oil in a frying pan and lightly fry the onion until golden brown.
Place cooked onion into a casserole dish.
Toss the venison in the seasoned flour and fry in the same oil until brown on each side and place in the casserole dish.

Make a roux in the frying pan with the remaining oil and flour.
Slowly add the water stirring all the time and pour into the casserole dish, add the vegetables and milk to the casserole and stir.
Cover and bake at 170°C (320°F) Mark 3 for about 3 hours or until the venison is tender.
Serve hot. Accompany with homemade redcurrant jelly, mashed potato, broad beans and carrot sticks garnished with sesame seeds.

Pork Cooked in Cider, Apples and Peppers

This dish incorporates the taste of the South West with use of cider and apples giving the pork a light refreshing taste.

600g (1 lb 6oz) ½" thick sliced lean pork
2 small onions or one large onion, chopped
4 eating apples, peeled, cored and sliced (not peeled if red)
1 red and 1 green pepper, deseeded and cut into strips
468ml (1 pint) medium sweet cider
300ml (½ pint) water
30ml (2 tbsp) sunflower oil
15ml (1 tbsp) plain flour
10ml (2 tsp) salt
10ml (2 tsp) ground pepper

Heat the sunflower oil in a frying pan and lightly fry the onion until golden brown and place in a casserole dish.
Lightly fry the pork in the same oil until slightly brown on each side and place in casserole dish.
Add the flour, salt and pepper to the remaining juices in the frying pan, stirring all the time, pour over the pork and onions.
Top up the liquid with the water.
Add the apples and peppers.
Cover casserole dish and bake at 170°C (320°F) Mark 3 for about 2 hours or until the pork in tender.
Serve hot. Accompany with carrot sticks, cauliflower garnished with paprika and new potatoes in butter.

Rabbit and Asparagus Pie

The Filling
1 Rabbit
1 carrot, chopped into 4 pieces
1 potato, chopped into 4 pieces
1 onion, chopped into 4 pieces
1 onion chopped
4 spears of asparagus
50g (2oz) margarine
25g (1oz) plain flour
300ml (½ pint) milk
1.1 litres (2 pints) water
Salt and pepper

450g (1 lb) shortcrust pastry
225g (8oz) plain flour
100g (4oz) margarine
75g (3oz) lard
15ml (3 tbsp) cold water, approximately
pinch of salt
1 beaten egg

The Filling
Place the rabbit and coarsely chopped carrot, potato and onion in a large pan with the water. Simmer for 1 hour until the meat begins to fall off the bones. Remove the meat from the bones.
The stock can be kept and used in a rabbit casserole.
Melt the margarine in a saucepan and lightly fry the chopped onion until soft but not brown. Add the flour to make an onion roux. Heating gently and stirring all the time, gradually add the milk to make an onion sauce.
Season to taste.
Remove from the heat and allow to cool before stirring in the rabbit meat.
Meanwhile make the pastry;
Put the flour, salt, margarine and lard into a bowl. Rub the margarine and lard into the flour until it resembles breadcrumbs. Add the water and mix to form a dough.
Roll out ½ the pastry and line a pie dish, prick with a fork.
Half cook the lining in a hot oven for 10 minutes.
Fill the cooked pie with the rabbit mixture and arrange the asparagus on top.
Using the remaining pastry make a pie lid and garnish with pastry leaves.
Make a 'X' in the middle of the pastry lid and brush with beaten egg.
Bake at 200°C (400°F) Mark 6 for 20 minutes or until golden brown.

West African Groundnut Stew
(from Antony's place of birth)

800g (2lbs) lean cubed stewing beef.
400g (1 lb) tin chopped tinned tomatoes
100g (¼ lb) tomato paste
2 onions, chopped
3 bay leaves
600ml (1 pt) water
A pinch or two of salt and pepper
45ml (3 tbsp) sunflower oil
200g (½ lb) jar smooth peanut butter
30ml (2 tbsp) lemon juice.

Lightly fry the onions until they turn golden brown and turn into a casserole dish. Lightly fry the stewing beef in the same oil and turn into the casserole dish. Add the chopped tomatoes, tomato paste, water, bay leaves and seasoning.
Cover and bake in a pre-heated oven at 170°C (325°F) Mark 3 for about 2 hours until the meat is tender.
Remove from oven to stir in the peanut butter and lemon juice and more water if necessary.
Return to oven for a further 30 minutes cooking time.
Serve accompanied with rice and a selection of the following, poppadoms, sultanas, sliced bananas, desiccated coconut, chopped boiled eggs, mango chutney, lime pickle, yogurt and cucumber, tabasco, nans and chapattis. Serves 6.

Whortleberry Cream

An easy, healthy dish with the added attraction that outside the whortleberry season it can be adapted to any soft fruit, such as rhubarb, raspberries and bananas.

450g (1 lb) stewed whortleberries
225g (8 oz) cottage cheese
100g (4 oz) castor sugar
10g (½ oz) gelatine,
Toasted nuts, clotted cream

Blend, process or beat together the cottage cheese and the sugar, add the stewed fruit and continue beating until smooth. Add the gelatine, softened in

a little water (and for special occasions, 2 tablespoons of kirsch or other liqueur).
Pour into individual bowls and leave to set. Decorate with flaked. toasted nuts and serve with clotted cream. Serves 6.

Rum Baked Bananas

A quick, microwave, hot delicious dish

4 bananas, peeled and sliced
4 tablespoons of rum
4 tablespoons orange juice
50g (2 oz) crunched ginger biscuits
50g (2 oz) soft brown sugar

Place the sliced bananas in an attractive microwave proof dish
Mix the rum and orange juice together and pour over bananas
Mix the biscuits and sugar together and sprinkle over the bananas.
Cover the dish and microwave on high for 4 minutes.
Serve with single or clotted cream.
Serves 4.

Tarka Blackberry Cream Fudge

Living on the newly opened "Tarka Trail" route inspired this creation. The combination of the wild blackberries and apple based "Tarka Liqueur" with the Devon Clotted Cream were acquired from the otters name "Tarka", meaning "the little wanderer" in the Devon countryside.
Any soft fruit such as strawberries, raspberries, loganberries, redcurrants and blackcurrants can be substituted.
The liqueur can be omitted if wished or replaced by any other liqueur.

300g (12oz) Blackberries
30ml (2 tbsp) Tarka Liqueur
300ml (1/2 pint) whipping cream
225g (8 oz) natural yogurt
100g (4 oz) Devon Clotted Cream
50g (2oz) soft brown sugar

Place the blackberries and liqueur into a pan and heat until soft.
Allow to cool and place the blackberry mixture in a glass bowl.
Whip the whipping cream until stiff, add the natural yogurt and whip again.
Add the clotted cream and whip in quickly. Pour over the blackberry mixture.

Sprinkler the soft brown sugar on top of the cream mixture and stand in a fridge for 2 hours to allow the sugar to soak into the cream mixture.
Serve cold.

Plum Cheesecake with Elderberry Sauce.

This is alight creamy cheesecake, the plums give it a slight sour tang complimented by the elderberries' sweet fruity taste.
Other fruits such as stewed rhubarb, strawberries and raspberries can be used in this recipe.

Plum Cheesecake
300g (10 oz) plums
150ml (¼ pint) water
100g (4 oz) sugar
10ml (2 tsp) gelantine
30ml (2 tsp) water
200ml (7 oz) whipping cream
175g (6 oz) soft cream cheese
150g (5 oz) digestive biscuits, crushed
75g (3 oz) West Country butter

Elderberry Sauce
150g (5 oz) elderberries
50g (2 oz) sugar
150ml (¼ pint) water
5ml (1 tsp) arrowroot

Plum Cheesecake
To make the base melt the butter in a pan and stir in the crushed biscuits.
Press into a 170mm (7") loose bottom flan tin.
Stew the plums with the water and sugar until the pips can be easily removed.
Mix the gelantine and 30ml of water in a small bowl and stand for 1 minute then microwave on high for 50 seconds, stir into the stewed plums and leave to cool but not to set.
Whip the whipping cream until stiff. Add the soft cream cheese and whip in quickly. Fold the stewed plum mixture to the cream and cheese and pour over biscuit base. Place in the fridge to set for 1 hour.
Garnish with sprigs of elderberries.

Elderberry Sauce
Place the elderberries, sugar and water in a pan and boil until
the sugar is dissolved.

Mix the arrowroot with a little water until it forms a smooth runny paste and to the elderberry mixture stirring all the time over a gentle heat.
Allow to cool. Serve cold with the plum cheesecake.

Elderflower Fritters with a Fudge Sauce

Elderflower Fritters
12 clusters of elderflowers
25g (1 oz) plain flour
1 egg, separated
150ml (¼ pint) milk
300ml (½ pint) sunflower oil
Pinch of salt

Fudge Sauce
50g (2 oz) soft brown sugar
50g (2 oz) sugar
75g (3 oz) clotted cream
15ml (1 tbsp) milk
5ml (1 tsp) vanilla essence

Finally, a recipe for "Crunch", which we sometimes offer to guests on their arrival, instead of Jackie's sponge cake. It is a South African recipe learned from a missionary in Malawi.

Crunch

¼ kilo block margarine
1 cup (large) sugar
1 heaped tbs golden syrup

Melt the above in a saucepan and and then stir in
1 tsp bicarb. soda
2 cups (large) plain flour
1 cup (large) desiccated coconut
1½ cups (large) oats

Bake this mixture in a greased baking tin (approx. 10" x 20") in a moderate oven until golden brown and cut into squares whilst hot.

Appendix 9. FARMHOUSE PLANS FROM 1520

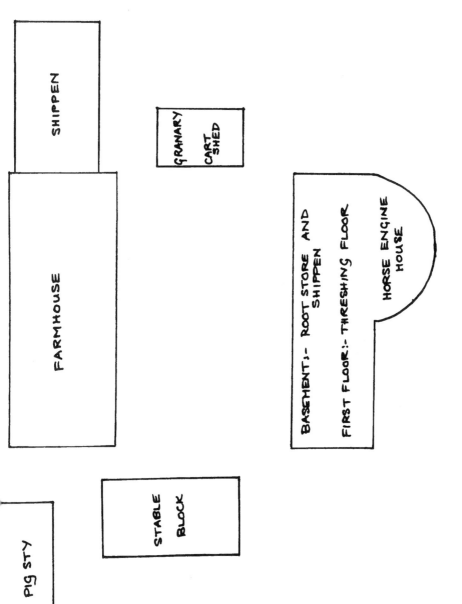

SHIPPEN

GRANARY

CART SHED

FARMHOUSE

BASEMENT:- ROOT STORE AND SHIPPEN

FIRST FLOOR:- THRESHING FLOOR

HORSE ENGINE HOUSE

STABLE BLOCK

PIG STY

The Farm Buildings in the 19th Century

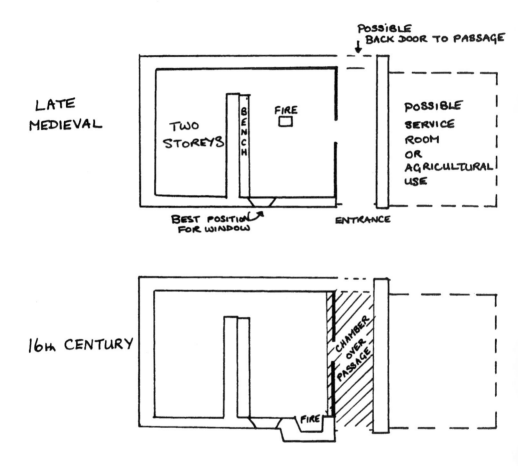

LATE MEDIEVAL

POSSIBLE ↓ BACK DOOR TO PASSAGE

TWO STOREYS

B E N C H

FIRE

POSSIBLE SERVICE ROOM OR AGRICULTURAL USE

BEST POSITION FOR WINDOW

ENTRANCE

16th CENTURY

CHAMBER OVER PASSAGE

FIRE

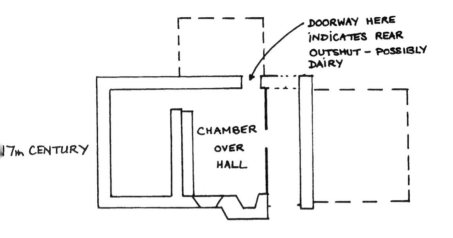

DOORWAY HERE
INDICATES REAR
OUTSHUT — POSSIBLY
DAIRY

17th CENTURY

CHAMBER
OVER
HALL

LATE 18th CENTURY

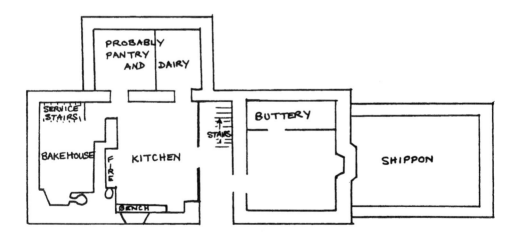

PROBABLY
PANTRY
AND DAIRY

SERVICE
STAIRS

BAKEHOUSE

FIRE

KITCHEN

BENCH

STAIRS

BUTTERY

SHIPPON

ACKNOWLEDGEMENTS

I have received much help in editing this book and wish to thank those who contributed chapters, John Huxtable, Lady Margaret Fortescue, Mary Camerom for the first ones and John Thorp for his professional research. Doreen Ridd's chapters are the result of many hours of work as are the later chapters by my daughter-in-law, Jackie who has also drawn the careful maps and illustrations and designed the cover; special thanks go to them.

"Huxtable" would not, however, have been finally published had it not been for Dr. James Porterfield whose computer expertise enabled Mary Cameron, who did most of the word processing, and myself to edit this book and to prepare it for the printers.

The initial impetus was North Devon Community Publications whose members provided help and advice. Thanks also go to the North Devon Records Office (and to the Exeter Record Office) who helped with the original research as did Paul Williams of the North Devon College.

NOTES

NOTES

NOTES

NOTES